AF576010

Ionic Liquids and Deep Eutectic Solvents for Application in Pharmaceutics

Ionic Liquids and Deep Eutectic Solvents for Application in Pharmaceutics

Editors

Miguel M. Santos
Luís C. Branco

MDPI • Basel • Beijing • Wuhan • Barcelona • Belgrade • Manchester • Tokyo • Cluj • Tianjin

Editors
Miguel M. Santos
LAQV-REQUIMTE,
Departamento de Química,
Faculdade de Ciências e
Tecnologia, Universidade Nova
de Lisboa
Portugal

Luís C. Branco
LAQV-REQUIMTE,
Departamento de Química,
Faculdade de Ciências e
Tecnologia, Universidade Nova
de Lisboa
Portugal

Editorial Office
MDPI
St. Alban-Anlage 66
4052 Basel, Switzerland

This is a reprint of articles from the Special Issue published online in the open access journal *Pharmaceutics* (ISSN 1999-4923) (available at: https://www.mdpi.com/journal/pharmaceutics/special_issues/ILs_DES).

For citation purposes, cite each article independently as indicated on the article page online and as indicated below:

LastName, A.A.; LastName, B.B.; LastName, C.C. Article Title. *Journal Name* **Year**, *Volume Number*, Page Range.

ISBN 978-3-0365-0056-0 (Hbk)
ISBN 978-3-0365-0057-7 (PDF)

Contents

About the Editors

Miguel M. Santos (PhD) received his PhD degree in Chemistry in 2013 from the University of Aveiro. His thesis focused on the supramolecular recognition of organic and inorganic anions by azacalix[2]arene[2]triazine macrocycles and isoftalamides, under the supervision of Professor Vitor Félix (UA) and Dr. Cristina Moiteiro (FCUL). He is currently a Researcher at the Cultural Heritage and Responsive Materials Group from LAQV-REQUIMTE focusing on the development of Organic Salts and Ionic Liquids from Active Pharmaceutical Ingredients as highly effective ionic formulations of drugs.

Luís C. Branco (PhD) has a Chemistry Degree (FCUL, 1998) and a PhD in Organic Chemistry (May 2006; FCT/UNL) under the supervision of Prof. Carlos Afonso; a Postdoctoral position (June 2006 to March 2008; IST/UTL and Cambridge, UK) in Supramolecular Chemistry, Assistant Research Position in REQUIMTE, FCT/UNL (2008–2013) and then Principal Research position at the same Institution (2014–2019). In December 2019, he was appointed Assistant Professor at FCT/UNL. His main scientific interests are focused on the development of Sustainable Chemistry and Applied Functional Materials including ionic system-based ionic liquids and eutectic solvents for sustainability.

Preface to "Ionic Liquids and Deep Eutectic Solvents for Application in Pharmaceutics"

Over the last few decades, ionic liquids (ILs) and deep eutectic solvents (DES) have been studied academically throughout many fields of chemical and biological research, including pharmaceutical sciences, due to their highly tunable physical, chemical, and physicochemical properties. In pharmaceutics, ILs have been studied as alternative solvents for the preparation, purification, and crystallization of active pharmaceutical ingredients (APIs). However, the tailorable properties of ILs and DES render them distinct interactions with cellular membranes and organelles, enabling a myriad of applications ranging from bactericidal agents against pathological microorganisms to innovative materials for transdermal drug and protein delivery. ILs and DES have also found use in the manipulation of naturally occurring matrices such as polysaccharides, yielding outstanding materials suitable for tissue regeneration and wound healing, as well as gene and drug delivery. Furthermore, there has been striking evidence over the last decade that the formation of ILs by a combination of ionizable APIs with biocompatible organic counter-ions (API-ILs) is a very capable instrument to enhance the bioavailability of poorly water-soluble and/or lipid-soluble drugs and reduce or even eliminate polymorphism, yielding more effective formulations of commercial drugs.

This Special Issue collected emergent discoveries on the application of Ionic Liquids and Deep Eutectic Solvents in pharmaceutical sciences. Six original papers and one review article were published.

Miguel M. Santos, Luís C. Branco
Editors

Editorial

Ionic Liquids and Deep Eutectic Solvents for Application in Pharmaceutics

Miguel M. Santos * and Luís C. Branco *

LAQV-REQUIMTE, Departamento de Química, Faculdade de Ciências e Tecnologia, Universidade Nova de Lisboa, 2829-516 Caparica, Portugal
* Correspondence: miguelmsantos@fct.unl.pt (M.M.S.); l.branco@fct.unl.pt (L.C.B.)

Received: 16 September 2020; Accepted: 22 September 2020; Published: 23 September 2020

Over the last few decades, Ionic Liquids (ILs) and Deep Eutectic Solvents (DES) have been studied academically throughout many fields of chemical and biological research, including pharmaceutical sciences, due to their highly tunable physical, chemical and physicochemical properties. In pharmaceutics, ILs have been studied as alternative solvents for the preparation, purification and crystallization of active pharmaceutical ingredients (APIs). However, the tailorable properties of ILs and DES render them distinct interactions with cellular membranes and organelles, enabling a myriad of applications ranging from bactericidal agents against pathological microorganisms to innovative materials for transdermal drug and protein delivery. ILs and DES have also found use in the manipulation of naturally occurring matrices such as polysaccharides, yielding outstanding materials suitable for tissue regeneration and wound healing, as well as gene and drug delivery. Furthermore, the formation of ILs by a combination of ionizable APIs with biocompatible organic counter-ions (API-ILs) has shown striking evidence over the last decade as a very capable instrument to enhance the bioavailability of poorly water and/or lipid-soluble drugs, in addition to reduce or even eliminate polymorphism, yielding more effective formulations of commercial drugs.

The goal of this Special Issue was to gather emergent discoveries on the application of Ionic Liquids and Deep Eutectic Solvents in pharmaceutical sciences. Six original papers and one review article were published.

The publication by Umerska & Tajber et al. [1] focused on anticrystal engineering of ketoprofen by combining it with benzocaine, procaine and tetracaine local anesthetics as Deep Eutectic Mixtures (DEMs). By melting and quench cooling a neat mechanical mixture of the components in discrete proportions, supercooled liquids were obtained for the vast majority of the prepared ketoprofen-based DEMs. A detailed thermal study was performed for all mixtures, which showed that the ketoprofen-procaine systems showed the most promising glass-forming ability of the three, with none of the components crystallizing when cooled or heated. In fact, this system was the only one that exhibited ionization, though partial, of ketoprofen's carboxylic acid moiety by infrared spectroscopy and molecular modeling.

Three papers on the development of Organic Salts and Ionic Liquids from three different families of Active Pharmaceutical Ingredients (API-OSILs) with enhanced in vitro therapeutic activity were also included in this Special Issue [2–4]. On all papers, NMR and infrared spectroscopic studies as well as elemental analysis confirmed the formation of the highly bioavailable API-OSILs.

In more detail, Teixeira, Santos & Branco and co-workers [2] reported the combination of the anti-bone resorption drug alendronic acid (ALN) with four different cations in 1:1 and 2:1 proportions by deprotonation of one or two of the ALN phosphonate groups, respectively. In general, the monoanionic ALN-OSILs tended to be less toxic to healthy cells as well as to breast, lung and bone (osteosarcoma) cell lines than the dianionic ones. All ALN-OSILs displayed lower antitumor activity than the standard paclitaxel, however in several cases a distinction between healthy and cancer cells was observed. In this regard, the authors pointed towards [C_2OHMIM][ALN] as the most promising ionic formulation of the set due to its efficiency towards lung cancer and osteosarcoma cell lines and very low toxicity towards fibroblasts.

By following one previously published methodology for the preparation of ampicillin-based OSILs [5], Ferraz & Branco and co-workers [3] reported the preparation of amoxicillin and penicillin antibiotics as OSILs bearing hydrolyzed β-lactam moieties (respectively *seco*-AMX- and *seco*-Pen-OSILs). Despite this shortcoming, on one hand, two of the hydrolyzed *seco*-Pen-OSILs showed up to a 100-fold increase in antimicrobial activity in vitro against sensitive *S. aureus* ([C_2OHMIM][*seco*-Pen]) and *E. coli* ([TEA][*seco*-Pen]) strains compared to parent non-hydrolyzed antibiotics. The authors confirmed that the hydrolyzed version of the parent antibiotics, as Na or K salts, did not affect these bacteria; hence, there must be a distinct underlying mechanism for the antimicrobial activity of these *seco*-Pen-OSILs. On the other hand, the combination of both hydrolyzed antibiotics with [Choline], [$P_{6,6,6,14}$] and [C_{16}Py] cations efficiently inhibited the growth of resistant *E. coli* and MRSA strains in vitro. In fact, [C_{16}Py][*seco*-AMX] recorded an increase of at least 1000-fold in comparison with the original antibiotic bearing an intact or hydrolyzed β-lactam moiety.

Santos & Branco and co-workers [4] reported the preparation of twelve new ciprofloxacin- and norfloxacin-based OSILs by combination with six different organic cations. The obtained isomorphous salts and ionic liquids were found to be highly water-soluble as well as non-toxic to mouse fibroblasts. The in vitro antimicrobial activity studies performed on sensitive pathogenic *K. pneumoniae* and *S. aureus* strains, as well as commensal *B. subtilis*, revealed that it is possible to develop fluoroquinolone-based OSILs with very distinct activities, depending on the suitable combinations with organic cations. For example, the combination of [C_{16}Py] cation with ciprofloxacin and norfloxacin selectively inhibited the growth of *K. pneumoniae* (20-fold) and *S. aureus* (11-fold), respectively, with a reduced effect on the other strains. On the other hand, the combinations of ciprofloxacin with [EMIM] or [C_2OHDMIM] cations are also promising, as they have enhanced activity against both pathogenic bacteria strains.

Two additional research papers on the development of lipid-based formulations (LBFs) of water-insoluble drugs by means of Ionic Liquids were also included in this Special Issue. The report by Tay & Porter et al. [6] focused on the dissolution profile of lumefantrine combined with docusate and dodecyl sulfate in nine different LBFs. Upon spectroscopic and physicochemical characterization of the two ILs, solubility studies showed particularly promising loadings with the docusate IL in four of the LBFs. Type III formulations were able to maintain solubilization and supersaturation of the drugs in vitro, which consequently prompted a 35-fold plasma exposure in vivo than with the remaining tested formulations. Moreover, in comparison with lipid and aqueous suspensions of the drug's free base, this formulation achieved respectively 10 and 50 times more oral bioavailability.

On the other hand, Islam & Goto et al. [7] prepared promising Ionic Liquid-in-oil microemulsions (IL/O ME) for the transdermal delivery of acyclovir. The MEs were composed of hydrophilic ILs, which were based on the combination of choline with formate, lactate and propionate, and choline oleate as the surfactant IL in combination with Span-20. The highly stable nano-sized droplets of ME allowed up to 7.7 mg/mL drug loadings (with the propionate salt) and significantly enhanced drug permeation across pig skin models by reduction of the skin barrier function with very high biocompatibility.

Transdermal drug delivery systems assisted by Ionic Liquids were also the focus of the review article by Sidat & Pillay et al. [8]. The physical, chemical and physicochemical properties of the ILs are discussed according to their lipophilic and hydrophilic behavior, as well as their possible interactions with biomolecules and biological membranes. The authors point out several examples in which ILs are used as skin permeation enhancers in both seldom or synergistic combinatorial use with striking advantages over standard systems. Moreover, the dissolution and complexation of ILs with drugs, as well as the preparation of ILs as Active Pharmaceutical Ingredients for assembly of transdermal drug delivery systems, have been revised.

Conflicts of Interest: The authors declare no conflict of interest.

References

1. Umerska, A.; Bialek, K.; Zotova, J.; Skotnicki, M.; Tajber, L. Anticrystal Engineering of Ketoprofen and Ester Local Anesthetics: Ionic Liquids or Deep Eutectic Mixtures? *Pharmaceutics* **2020**, *12*, 368. [CrossRef] [PubMed]
2. Teixeira, S.; Santos, M.M.; Fernandes, M.H.; Costa-Rodrigues, J.; Branco, L.C. Alendronic Acid as Ionic Liquid: New Perspective on Osteosarcoma. *Pharmaceutics* **2020**, *12*, 293. [CrossRef] [PubMed]
3. Ferraz, R.; Silva, D.; Dias, A.R.; Dias, V.; Santos, M.M.; Pinheiro, L.; Prudêncio, C.; Noronha, J.P.; Petrovski, Ž.; Branco, L.C. Synthesis and Antibacterial Activity of Ionic Liquids and Organic Salts Based on Penicillin G and Amoxicillin hydrolysate Derivatives against Resistant Bacteria. *Pharmaceutics* **2020**, *12*, 221.
4. Santos, M.M.; Alves, C.; Silva, J.; Florindo, C.; Costa, A.; Petrovski, Ž.; Marrucho, I.M.; Pedrosa, R.; Branco, L.C. Antimicrobial Activities of Highly Bioavailable Organic Salts and Ionic Liquids from Fluoroquinolones. *Pharmaceutics* **2020**, *12*, 694. [CrossRef] [PubMed]
5. Ferraz, R.; Teixeira, V.; Rodrigues, D.; Fernandes, R.; Prudêncio, C.; Noronha, J.P.; Petrovski, Ž.; Branco, L.C. Antibacterial activity of Ionic Liquids based on ampicillin against resistant bacteria. *RSC Adv.* **2014**, *4*, 4301–4307.
6. Tay, E.; Nguyen, T.-H.; Ford, L.; Williams, H.D.; Benameur, H.; Scammells, P.J.; Porter, C.J.H. Ionic Liquid Forms of the Antimalarial Lumefantrine in Combination with LFCS Type IIIB Lipid-Based Formulations Preferentially Increase Lipid Solubility, In Vitro Solubilization Behavior and In Vivo Exposure. *Pharmaceutics* **2020**, *12*, 17. [CrossRef] [PubMed]
7. Islam, M.R.; Chowdhury, M.R.; Wakabayashi, R.; Kamiya, N.; Moniruzzaman, M.; Goto, M. Ionic Liquid-In-Oil Microemulsions Prepared with Biocompatible Choline Carboxylic Acids for Improving the Transdermal Delivery of a Sparingly Soluble Drug. *Pharmaceutics* **2020**, *12*, 392. [CrossRef] [PubMed]
8. Sidat, Z.; Marimuthu, T.; Kumar, P.; du Toit, L.C.; Kondiah, P.P.D.; Choonara, Y.E.; Pillay, V. Ionic Liquids as Potential and Synergistic Permeation Enhancers for Transdermal Drug Delivery. *Pharmaceutics* **2019**, *11*, 96.

Article

Anticrystal Engineering of Ketoprofen and Ester Local Anesthetics: Ionic Liquids or Deep Eutectic Mixtures?

Anita Umerska [1], Klaudia Bialek [1], Julija Zotova [1], Marcin Skotnicki [1,2] and Lidia Tajber [1,*]

[1] School of Pharmacy and Pharmaceutical Sciences, Trinity College Dublin, College Green, Dublin 2, Ireland; umerskaa@tcd.ie (A.U.); bialekk@tcd.ie (K.B.); zotovaj@tcd.ie (J.Z.); marcskot@ump.edu.pl (M.S.)

[2] Department of Pharmaceutical Technology, Poznań University of Medical Sciences, 60-780 Poznań, Poland

* Correspondence: ltajber@tcd.ie; Tel.: +353-1-896-2787

Received: 30 March 2020; Accepted: 14 April 2020; Published: 17 April 2020 

Abstract: Ionic liquids (ILs) and deep eutectic mixtures (DEMs) are potential solutions to the problems of low solubility, polymorphism, and low bioavailability of drugs. The aim of this work was to develop and investigate ketoprofen (KET)-based ILs/DEMs containing an ester local anesthetic (LA): benzocaine (BEN), procaine (PRO) and tetracaine (TET) as the second component. ILs/DEMs were prepared via a mechanosynthetic process that involved the mixing of KET with an LA in a range of molar ratios and applying a thermal treatment. After heating above the melting point and quench cooling, the formation of supercooled liquids with Tgs that were dependent on the composition was observed for all KET-LA mixtures with exception of that containing 95 mol% of BEN. The KET-LA mixtures containing either ≥ 60 mol% BEN or 95 mol% of TET showed crystallization to BEN and TET, respectively, during either cooling or second heating. KET decreased the crystallization tendency of BEN and TET and increased their glass-forming ability. The KET-PRO systems showed good glass-forming ability and did not crystallize either during the cooling or during the second heating cycle irrespective of the composition. Infrared spectroscopy and molecular modeling indicated that KET and LAs formed DEMs, but in the KET-PRO systems small quantities of carboxylate anions were present.

Keywords: ionic liquids; deep eutectic mixtures; ketoprofen; non-steroidal anti-inflammatory drugs (NSAIDs); local anesthetics; procaine; benzocaine; tetracaine; anticrystal engineering; mechanochemistry

---

1. Introduction

The pharmaceutical industry relies on solid, predominantly crystalline, forms of active pharmaceutical ingredients (APIs) [1]. Crystalline APIs, including multicomponent systems such as salts and co-crystals, exhibit polymorphism that could negatively influence the solubility and bioavailability of the API [2]. Ionic liquids (ILs) have recently been identified as a promising approach to overcome the above-mentioned problems [3–5]. The IL form can improve the API performance by providing controlled solubility and drug delivery [6,7]. An IL is a multicomponent, one-phase system composed of ionized species with either a melting point (Mp) or a glass transition temperature (Tg) below 100 °C [6]. However, it has been shown that not only the proton transfer, but also strong hydrogen bond formation can be the driving force for the liquefaction of solid APIs, leading to the formation of deep eutectic mixtures (DEMs) [4,8,9]. For instance, it has been demonstrated that lidocaine ibuprofenate, generally considered as an IL [10], is in fact a DEM due to the low degree of ionization/proton transfer [2]. APIs are transformed into ILs by combining them with appropriate counterions. These counterions should ideally be monovalent, with a minimal number of potential H-bonds between molecules, asymmetric, with soft electron density, bulky with voluminous side chains causing steric inhibition crystallization among the salt components [6]. The primary aim of the counterion is to lower the Mp of the salt, yielding a low-melting point phase. However, these counterions can modify the properties such as solubility, lipophilicity, partition coefficient, thereby affecting the API transport via biological membranes, its absorption and/or other pharmacokinetic parameters [1–4,10].

Interestingly, some APIs satisfy the above-mentioned IL counterion requirements [6]. The dual functional ILs (i.e., ILs composed of two APIs) offer the advantage of lowering of the Mp of API, in addition to providing a second biological function [2–4]. The potential clinical benefits and the increasing research and development cost of each new molecular entity have attracted interest in developing drug-drug combinations [11,12]. Hence, the possibility of assembly of drug combination in the form of ILs or DEMs, whereby two API molecules are associated together via reversible intermolecular interactions, is a very attractive approach, especially for pharmaceutical combination products. For instance, non-steroidal anti-inflammatory drugs (NSAIDs) and local anesthetics (LAs) could create a successful dual drug combination because both are widely used in the treatment of acute and chronic pain, as non-opioid strategies for pain management [13]. NSAIDs are generally acids, whereas LAs are bases, hence there is a potential for a proton exchange. There have been attempts to synthesize NSAID-LA ILs, with ibuprofen being the most commonly used NSAID and lidocaine, the most commonly used LA [2,10,14]. Ketoprofen (KET) is another very interesting and promising NSAID candidate to form ILs and information on KET-based ILs is scarce, with only tetrabutylphosphonium [7] and choline [15] used as the counterions. To the best of authors' knowledge, there is no reports describing KET ILs or DEMs with LAs. Among LAs from ester group, there are two interesting APIs with low melting point that meet the IL counterion requirements: tetracaine (TET) and procaine (PRO). Another important compound from the ester LAs group, benzocaine (BEN), has a high crystallization tendency [16] and the presence only a weak base group (aromatic amine), thus providing a challenge for the IL formation.

The great majority of studies conducted thus far have focused on equimolar mixtures of the NSAID and LA [5,10,17], whereas only three NSAID-LA examples of complete phase diagrams, i.e., lidocaine-ibuprofen [2], lidocaine-indomethacin [18], and lidocaine-naproxen [19] have been described. However, due to the crucial importance of non-ionic interactions in DEMs, it is important to consider a wide variety of component ratios in addition to the equimolar mixture.

ILs are usually synthesized by metathesis reactions, which involve dissolving of the salts of the acid and the base components of ILs in a suitable solvent such as water, methanol, ethanol, within which the IL readily forms [6]. However, metathesis has considerable disadvantages such as the need for product purification, removal of counterions such as sodium, chloride, use of organic solvents and compromised product yield. Alternatively, the solvent-free process that involves co-melting of free forms of the acid and base components was used by Cherukuvada and Nangia [20] to obtain the ILs of ethambutol such as ethambutol adipate.

The aim of this work was to develop and characterize KET-ester LAs ILs/DEMs at various mixing ratios and construct the phase diagrams to inform selection of the compositions with pharmaceutical relevance. Heating was employed to obtain ILs/DEMs via mechanosynthesis and to understand the thermal aspects of KET-LA mixing. The glass-forming ability and/or crystallization tendency of the studied mixtures was also investigated, and solid-state phase identification conducted. Infrared spectroscopy was the method employed to investigate the nature of the interaction between the components (i.e., H-bonding versus proton transfer) and to ultimately determine whether they form a DEM or an IL. To support the experimental efforts, molecular modelling was employed to understand intermolecular interactions in the KET-LA mixtures.

2. Materials and Methods

2.1. Materials

Ketoprofen (KET) was purchased from Fluorochem Ltd., (Hadfield, UK). Benzocaine (BEN), procaine (PRO) and tetracaine (TET) were purchased from TCI (Tokyo Chemical Industry UK Ltd., Oxford, UK). All chemicals were used as supplied and had purity of at least 98%.

2.2. Sample Preparation

The binary mixtures were prepared via mixing of an acid (KET) with a base (an LA). A quantity of 200 mg of each powder mixtures containing KET and each of the LA bases was prepared at a range of KET/LA mol% between 5 and 95. The constituents were accurately weighed using an MT5 Mettler Toledo microbalance (Greifensee, Switzerland) and ground in an agate mortar using an agate pestle until a homogenous powder mixture was obtained.

2.3. Thermogravimetric Analysis (TGA)

TGA was performed to evaluate the thermal stability of samples to make sure that they do not decompose during heating. TGA was carried out on the powders using a Mettler Toledo TG50 measuring module coupled to a Mettler Toledo MT5 balance [21]. Approximately 8–10 mg samples were analyzed ($n = 3$) in open aluminum pans, using nitrogen as the purge gas with a flow rate of 40 mL/min. Samples were heated from 25 to 250 °C at a rate of 10 °C/min. Mettler Toledo STAR^e software (version 6.10) was used to determine the mass loss based on the slope of TGA trace. Samples were considered as stable at temperatures at which the mass loss was smaller than 5% of the initial sample weight.

2.4. Differential Scanning Calorimetry (DSC)

DSC was carried out using a Perkin Elmer Diamond DSC unit (Waltham, MA, USA) with a ULSP B.V. 130 cooling system (Ede, Netherlands). Nitrogen with a flow rate of 40 mL/min was employed as a purge gas that was controlled with a PerkinElmer Thermal Analysis Gas Station. The instrument was calibrated for temperature and heat flow using an indium calibration reference material (99.999%, transition point 156.60 °C) supplied by Perkin Elmer. Approximately 3–5 mg of accurately weighed powder (MT5 Mettler Toledo microbalance) were analyzed in 18 µL aluminum pans with covers. The samples were held at either 25 °C (KET-BEN) or 0 °C (KET-PRO and KET-TET) for 1 min in the DSC unit, then heated at a rate of 10 °C/min (=first heating), cooled at a fast, nominal rate of 300 °C/min, held at −60 °C for 5 min and reheated at a rate of 10 °C (=second heating). Thermal analysis was repeated at least three times for every sample. Samples referred to as quench cooled (QC) were obtained by first melting in situ in the DSC and then cooled at a nominal rate of 300 °C/min. Pyris software (version 9.01.0174) was used to analyze the thermograms.

2.5. Hyperscan DSC (HSDSC)

HSDSC measurements were carried out using a PerkinElmer Diamond DSC unit described in Section 2.4 using helium at a flow rate of 60 mL/min as the purge gas. Approximately 0.5 mg of accurately weighed powder (MT5 Mettler Toledo microbalance) was analyzed ($n = 3$) in 18 µL aluminum pans with covers. Samples were heated from 0 to 110 °C using heating rates of 50, 100 and 300 °C/min.

2.6. Powder X-ray Diffraction (PXRD)

Powder X-ray diffraction analysis, in duplicate for every sample studied, was performed with a Rigaku Miniflex II desktop X-ray diffractometer (Tokyo, Japan) set to 30 kV and 15 mA and equipped with a Haskris cooling unit (Elk Grove Village, IL, USA). Ni-filtered Cu Kα radiation ($\lambda = 1.5408$ Å) was used. The measurements were carried out in a range of 5–40 2theta degrees at a step size of 0.05°

per second at room temperature. The samples were front-loaded onto a low background silicon mount (Rigaku, Tokyo, Japan) [22].

2.7. Attenuated Total Reflection Fourier Transform Infrared Spectroscopy (ATR-FTIR)

ATR-FTIR experiments ($n = 2$) were performed using a PerkinElmer Spectrum 100 FTIR spectrometer (Shelton, CT, USA) equipped with a PerkinElmer universal ATR sampling accessory. A spectral range of 650–4000 cm^{-1}, resolution of 4 cm^{-1}, accumulation of 10 scans and data interval of 1 cm^{-1} were used. Spectrum software version 10.6.0 was used for background correction, normalization, and infrared spectra analysis.

2.8. Analysis of Thermal Data

Experimental phase diagrams were constructed using the thermal data obtained as described in Section 2.4. Theoretical phase diagrams of an ideal eutectic mixture, which implies the existence of complete insolubility between the two components at all concentrations, were also calculated. The theoretical phase diagrams were constructed using the Schröder van Laar equation (Equation (1)) [23]:

$$lnx_i = -\frac{\Delta H_{fi}}{R}\left(\frac{1}{T} - \frac{1}{T_{fi}}\right) \quad (1)$$

where x_i is the mole fraction of the component i at temperature T (in Kelvin), R is the gas constant (R = 8.314 J K^{-1} mol^{-1}), ΔH_{fi} is the molar enthalpy of fusion of component i, and T_{fi} is the melting temperature (onset) of pure component i (in Kelvin). ΔH_{fi} and T_{fi} were determined by DSC (Section 2.4).

The eutectic composition, i.e., the composition with the maximal ΔH of the eutectic peak, was determined for the KET-BEN mixtures using the Tammann plot, which represents ΔH of the eutectic peak as a function composition.

Experimental Tgs were compared with those calculated based on the knowledge about the properties of the pure components [24] using the Fox equation (Equation (2)):

$$\frac{1}{Tg_m} = \frac{w_1}{Tg_1} + \frac{w_2}{Tg_2} \quad (2)$$

where Tg is the glass transition temperature (midpoint) (in Kelvin degrees), w is the weight fraction concentration in the mixture and the subscripts denote component 1, component 2 and the mixture (m), respectively, [24]. Tg of pure components were determined by DSC (Section 2.4).

For the BEN- and TET-containing samples, for which melting was observed in the second heating cycle, the percentage of crystallinity was calculated using the following formula (Equation (3)) [25]:

$$\%\ crystallinity = \frac{\Delta H}{\Delta H_{ref}} \cdot 100\% \quad (3)$$

where ΔH is the enthalpy of melting obtained from DSC and corrected for the content of either BEN or TET and ΔH_{ref} is the enthalpy of melting of the reference, which was considered 100% crystalline (i.e., ΔH of BEN and TET from the second heating, 132.2 J/g and 116.4 J/g, respectively, as determined experimentally in this work).

2.9. Density Functional Theory (DFT) Calculations

Gaussian03 program was employed to optimize the structures of KET, PRO, TET and BEN [26]. The B3LYP/6-311++G(d,p) level of density functional theory (DFT) was used with no constraints on the geometry of molecules imposed. The dipole moment and charges were estimated using the CHelpG algorithm. A number of global electronic descriptors were calculated according to Koopmans' theorem [27]. The electron affinity (A) was expressed in terms of the HOMO (Highest Occupied Molecular Orbital) energy (E_{HOMO}) and ionization energy (I) as the LUMO (Lowest Unoccupied

Molecular Orbital) energy (E_{LUMO}). The following global reactivity descriptors were calculated: energy band gap (ΔE), absolute electron negativity (χ), absolute hardness (η) and electrophilicity index (ω) as per the equations published previously [9,28,29]. Wavefunction analysis was also conducted using Multiwfn 3.6 [30,31].

3. Results

3.1. Chemical Structure and Relevant Physicochemical Properties of Investigated Molecules

The relevant physicochemical properties of investigated molecules are shown in Table 1 and their structural formulas are depicted in Figure 1. KET belongs to the propionic acid class of NSAIDs, it is a weak acid due to the presence of a carboxylic acid group (pK_a = 3.88) and it bears negative charge at physiological pH (7.4). Each molecule of KET has one hydrogen donor and three hydrogen acceptors. TET (also known as amethocaine), PRO and BEN are local anesthetic drugs from the amino ester group. The structure of these local anesthetics consists of three components: a lipophilic part, an intermediate aliphatic chain and a hydrophilic (amine) part, with the ester-type linkage between the lipophilic part and the intermediate chain. The p-aminobenzoic group of ester-type LAs is divided into three subgroups based on the structural differences: procaine group, tetracaine group and benzocaine group. Hence, the investigated molecules are key representatives of each of these groups. PRO has the unsubstituted p-aminobenzoic group and tertiary amine as the hydrophilic group. TET also has the tertiary amine group as the hydrophilic group, but in contrast to PRO, it has a substituted p-aminobenzoic group. BEN is an ester of p-aminobenzoic acid with an unsubstituted amino group and it lacks the terminal tertiary amine group. Except for BEN, the pK_a of the tested LAs is greater than 8, so they are positively charged at pH 7.4. All LAs tested contain one hydrogen donor per molecule. TET and PRO contain three hydrogen acceptors per molecule, whereas BEN contains two hydrogen acceptors.

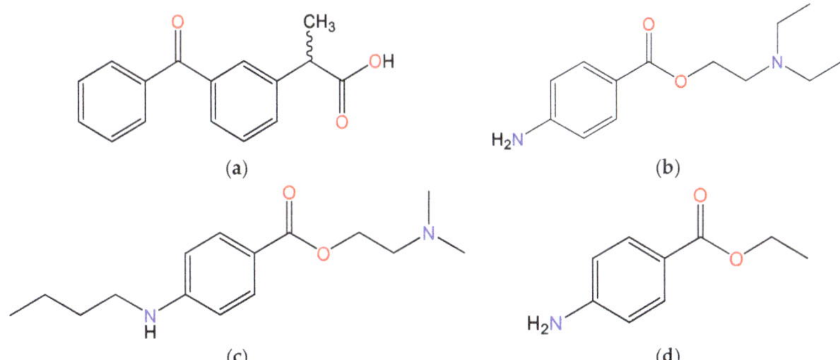

Figure 1. Chemical structures of the investigated molecules: (**a**) ketoprofen (KET), (**b**) procaine (PRO), (**c**) tetracaine (TET) and (**d**) benzocaine (BEN).

Table 1. Physicochemical properties of the investigated molecules (obtained from [32]).

Molecule	Ketoprofen	Tetracaine	Procaine	Benzocaine
Molecular weight	254.28 g/mol	264.37 g/mol	236.32 g/mol	165.19 g/mol
Melting point [1]	94.7 ± 0.4 °C	42.2 ± 0.0 °C	60.5 ± 0.1 °C	90.0 ± 0.1 °C
ΔH [1]	106.0 ± 1.3 J/g	144.2 ± 0.2 J/g	106.0 ± 0.5 J/g	129.3 ± 0.7 J/g
pK_a [2]	3.88	8.42	8.96	2.78
Physiological charge	−1	+1	+1	0
Hydrogen donor counts	1	1	1	1
Hydrogen acceptor counts	3	3	3	2

[1] Obtained experimentally in this work. [2] Strongest acidic for KET, protonated amine for LAs.

3.2. Ketoprofen-Procaine (KET-PRO) Systems

KET starting material in the first heating cycle showed a single melting endotherm with an onset at 94.7 ± 0.4 °C. PRO starting material in the first heating also exhibited a single melting endotherm (onset: 60.5 ± 0.1 °C). The heat treatment of the binary KET-PRO mixtures resulted in the formation of a eutectic phase (Figure 2a) and a phase transition to viscous liquids. The Mps of KET and PRO decreased, and the melting endotherms became broader corresponding to their decreasing content in the mixtures. DSC traces of the KET-PRO samples containing 30–80 mol% of KET showed a very broad endotherm with an onset at around 30 °C, which had a low enthalpy of transition. This event can be considered as melting of the eutectic phase. The experimental Mps of PRO in samples containing 5–20 mol% of KET agreed with the predicted values calculated using the Schröder van Laar equation Equation (1), whereas the Mps of KET were lower (Figure 2b). The experimental eutectic temperature was also lower than that determined from Equation (1), which was 46 °C. Therefore, the behavior of PRO and KET in KET-rich mixtures cannot be considered as that of an ideal system, where no strong intermolecular interactions take place. The calculated eutectic composition is 33 mol% of KET, but it was impossible to determine the experimental eutectic composition, due to an overlapping peak of PRO melting.

PXRD analysis of the KET-PRO physical mixtures prior to the heat treatment revealed the presence of crystalline substances with the peaks corresponding to both, PRO and KET starting materials (Figure 3a). The phase behavior of an equimolar KET-PRO mixture was further investigated by HSDSC (Figure 4a). When heated at a standard rate (10 °C/min), a very broad endotherm with an onset at 28.7 ± 0.4 °C was observed, which can be attributed to the eutectic melting followed by melting/dissolution of the excess of either KET or PRO. The onset of this eutectic endotherm moved from 40.8 ± 1.6 °C to 50.6 ± 1.5 °C when the heating rate increased from 50 °C/min to 300 °C/min and the peak became narrower. Moreover, in the equimolar mixture, at heating rates 50–300 °C/min, a second endotherm with an onset at approximately 92 °C appeared, which was attributed to KET. KET and PRO starting materials at heating rates 50–300 °C/min showed a single narrow melting endotherm with an onset at approximately 92 and 57 °C, respectively, independent on the heating rate (data not shown). Thus, the fast thermal treatment prevented the eutectic phase formation and a phase separation occurred, indicating that the eutectic was, indeed, composed of KET and PRO.

KET did not crystallize either during the fast cooling step or during the second heating (10 °C/min) as evidenced by the lack of the endothermic events and by a glass transition event (Tg) in the second heating cycle (midpoint: −2.7 ± 0.5 °C) (Figure 2c,d). Similarly, PRO did not crystallize either during cooling or during the second heating cycle and exhibited a Tg at −39.1 ± 0.4 °C. Quench cooled (QC) PRO and KET were disordered by PXRD (Figure 3a). In the phase diagram of the heat-treated KET-PRO samples (Figure 2c) only one phase transition was observed, i.e., the Tg, consistent with the disordered "halo" pattern recorded by PXRD (Figure 3a). The presence of one Tg in the binary mixtures implies mixing of the components at the molecular level. The Tg values, which were dependent on mixture composition (Figure 2d), indicate that heat-treated KET-PRO mixtures formed supercooled liquids. The experimental Tg values of samples containing at least 30 mol% of KET were larger than the predicted values calculated by the Fox equation (Equation (2)), thereby implying strong interactions between PRO and KET in the glassy state, which could involve proton transfer or at least strong H-bonds. KET-PRO supercooled liquids showed no tendency to crystallize when heated at 10 °C/min supported by the lack of crystallization exotherms and melting endotherms during the second heating. Moreover, slower heating and/or cooling rates (2 °C/min) did not induce crystallization.

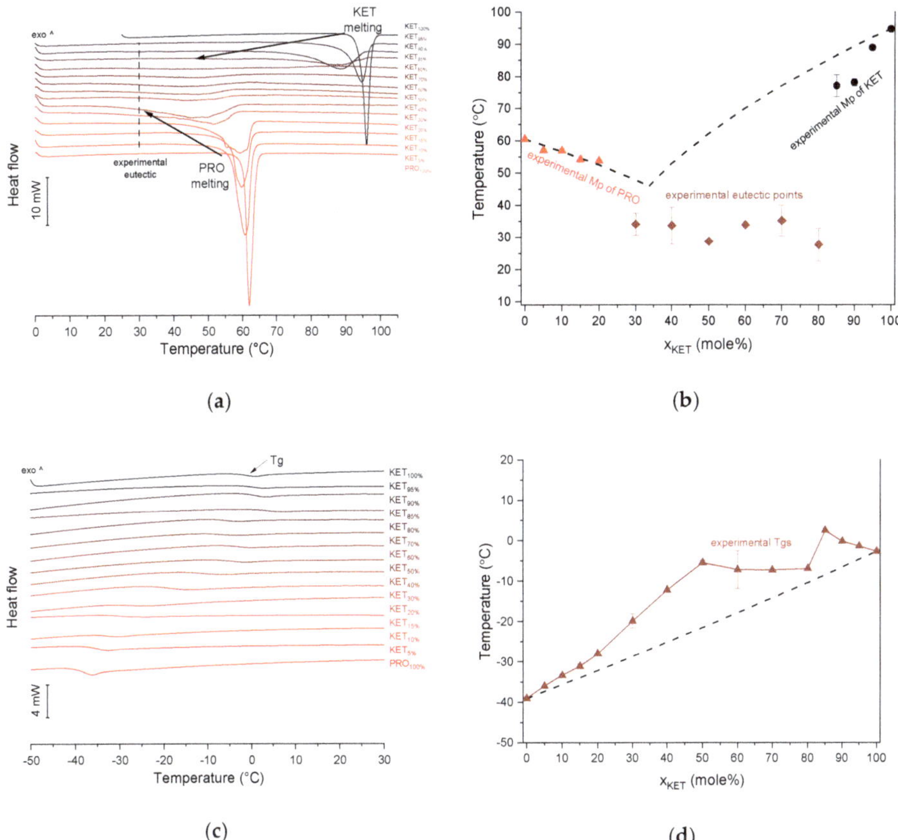

Figure 2. Thermal analysis of KET-PRO systems: (**a**) DSC thermograms from first heating; the broken line indicates the position of the eutectic peak while the arrows show the position of the melting peaks, (**b**) phase diagram based on thermal analysis of first DSC heating; the broken lines show the theoretical liquidus curves calculated using Equation (1), (**c**) DSC thermograms from second heating (the samples were first heated to 110 °C at 10 °C/min, quench cooled at a nominal cooling rate of 300 °C/min and reheated at 10 °C/min) (**d**) phase diagram based on thermal analysis of second DSC heating; the broken line shows the theoretical Tg values calculated using Equation (2). For plots (a) and (c) the subscript indicates the content of the named component in mole%.

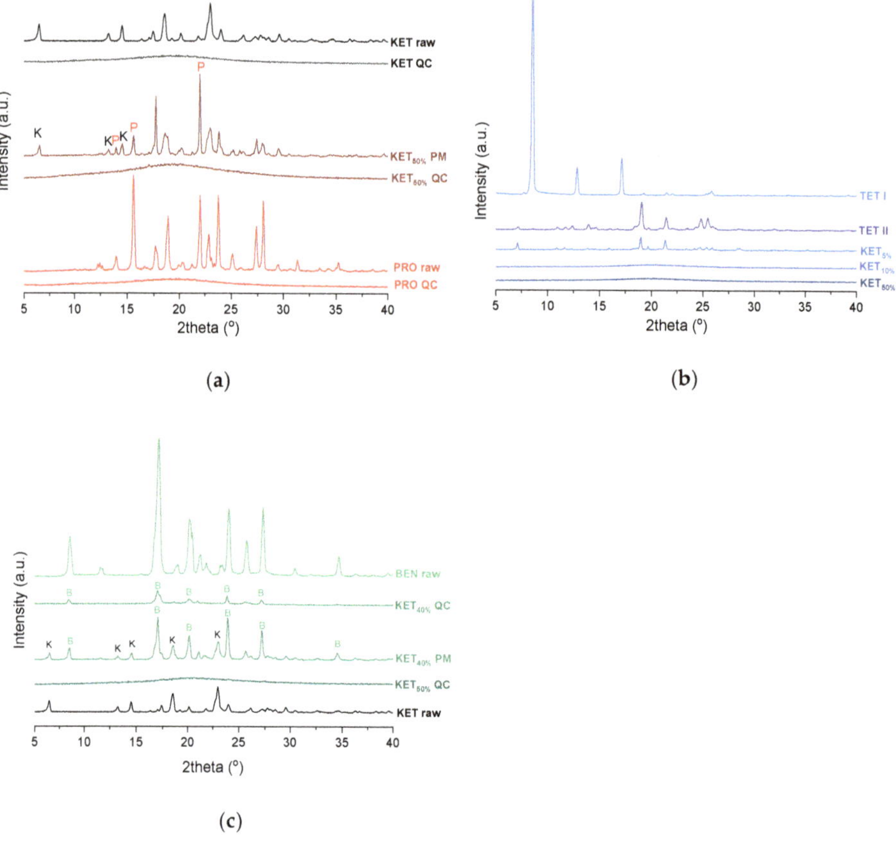

Figure 3. PXRD of: (**a**) KET, PRO and equimolar KET-PRO mixtures; (**b**) KET, TET and KET-TET mixtures; (**c**) KET, BEN and KET-BEN mixtures. QC-quench cooled, PM-physical mixture, raw-as supplied, TET I-TET polymorphic form I, TET II-TET polymorphic form II. The subscript indicates the KET content in mole%.

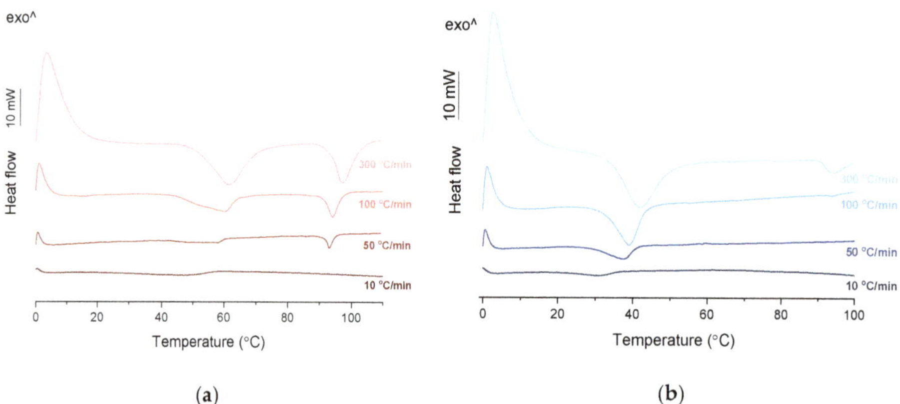

Figure 4. HSDSC thermograms of equimolar KET-LA powder mixtures heated at different rates: (**a**) KET-PRO and (**b**) KET-TET.

The infrared spectra of QC PRO, KET and KET-PRO mixtures are shown in Figure 5a. The band at 1695 cm^{-1} in the crystalline KET (starting material) can be ascribed to the C=O stretching of the carboxylic groups of KET molecules organized in dimers in the crystal lattice [33]. In the QC KET, which contained carboxylic acid monomers due to release of KET molecules from crystal lattice, this band shifted towards higher wavenumbers, giving a peak with the center at 1704 cm^{-1} and shoulder at 1737 cm^{-1}. The band at 1655 cm^{-1}, assigned to the stretching mode of the ketone group [33], was observed at 1656 cm^{-1} in QC KET. The broad peak in the range 2200–3400 cm^{-1} observed in both, KET starting material and QC KET can be assigned to the O–H stretching [33]. It overlaps with the alkane C–H stretching band at 2800–3000 cm^{-1} [34]. In the spectrum of crystalline PRO, two sharp peaks at 3463 and 3366 cm^{-1} can be ascribed to the N–H stretching bands [35]. The NH_2 groups in the crystal structure form intermolecular hydrogen bond with the ester C=O group [36]. The intensity of these N–H stretching bands decreased drastically in the QC PRO. The band at 1600 cm^{-1} can be attributed to the NH_2 scissoring mode [35]. The band at 1665 cm^{-1} in PRO starting material is probably the ester C=O stretching [35]. In the spectrum of QC PRO this band shifted towards higher wavenumbers (1689 cm^{-1}). The band at 1272 cm^{-1}, the most intensive band in the spectrum of PRO starting material, can be attributed to the ester C–O stretching [35]. The band at 1272 cm^{-1}, most likely of the C–O stretching, shifted towards lower wavenumbers (1267 cm^{-1}) in the spectrum of heated PRO. In the spectra of the QC KET-PRO mixtures, the doublet of N–H stretching bands shifted towards lower wavenumbers (3444 and 3361 cm^{-1} for equimolar mixture), but in the samples containing 70–90 mol% of KET it shifted back towards higher wavenumbers. The broad O–H stretching band was present in the sample containing 90 mol% of KET, but it was difficult to observe it in the sample containing 80 mol% of KET. The position of the N–H scissoring band (1600 cm^{-1}) did not change in QC KET-PRO mixtures, neither did the position of the ester C–O stretching band. The ketone C=O stretching band shifted slightly towards lower wavenumbers (1652 cm^{-1} in the equimolar mixture), and in the sample containing 10 mol% of KET it was no longer observed. In the case of the ester C=O band of PRO (1689 cm^{-1}) and carboxyl C=O band of KET (1704 cm^{-1}), the gradual change of one band into another was observed, in the equimolar sample the position of the 'hybrid' band was 1697 cm^{-1}. The shoulder of the KET carboxyl C=O monomers shifted towards lower wavenumbers in samples containing 90–70 mol% of KET and it was no longer observed in the sample containing 60 mol% of KET. Hence, the carboxyl group of KET is strongly involved in interactions with PRO molecules. The carboxylate group bands should appear at 1650–1550 cm^{-1} and 1400 cm^{-1}, arising from asymmetric and symmetric stretching, respectively, with the former that should be intensive and the latter weak. An increased absorbance in the region above 1550 cm^{-1} was observed in the spectra of QC KET-PRO samples. It was the most intensive in samples containing 60–70 mol% of KET. Hence, it is possible that a fraction of KET molecules is ionized in the supercooled mixtures and the proton is transferred probably to tertiary amine of PRO.

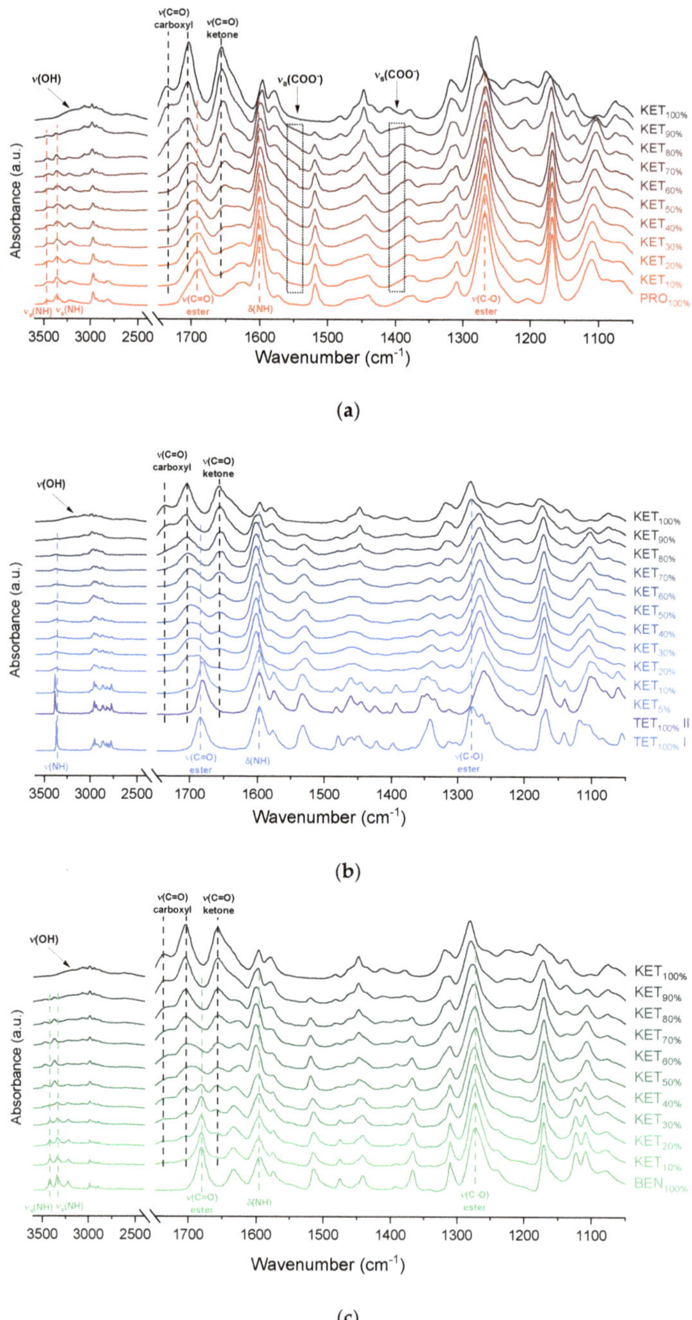

Figure 5. Infrared spectra of: (**a**) quench-cooled KET, PRO and KET-PRO mixtures; (**b**) quench-cooled KET, TET I, TET II and KET-PRO mixtures and (**c**) quench-cooled KET, BEN and KET-BEN mixtures. ν—stretching, ν_a—asymmetric stretching, ν_s—symmetric stretching and Δ—bending vibrations. The subscript indicates the content of the named component in mole%.

3.3. Ketoprofen-Tetracaine (KET-TET) Systems

TET showed a single melting endotherm (onset: 42.2 ± 0.0 °C) in the first heating (Figure 6a,b). The Mp of KET and TET decreased and the melting endotherms became broader corresponding to the decreasing quantity of the drugs in the sample. The KET-TET samples with 5–10 mol% of KET showed an Mp similar to that calculated using the Schröder van Laar equation, whereas all other experimental Mps were lower than those predicted for ideal systems. Similar to the KET-PRO mixtures, the formation of a eutectic phase was observed. A decrease in Mp in mixtures containing 15–20 mol% of KET may be explained by accelerated dissolution of TET crystals in the presence of the liquid phase. The eutectic point is probably close to that observed for 30–70 mol% of KET mixtures with an onset of approximately 21–23 °C, hence it is lower than the value calculated for an ideal system (35.8 °C). The theoretical eutectic composition determined from Equation (1) was 26 mol% of KET, but it was impossible to determine the experimental composition, due to overlapping peak of the TET melting endotherm (Figure 6a). PXRD revealed that the equimolar sample (unheated) displayed a pattern corresponding to the mixture of crystalline TET and KET (Figure 3b). To better understand the phase behavior of KET-TET systems, the equimolar physical mixture was heated at different rates. At 300 and 100 °C/min two melting endotherms were observed; the first one that appeared at approximately 33.7 ± 0.9 °C, and the second one had an onset above 90 °C (melting of KET) (Figure 4b). When the heating rate was reduced to 50 °C/min, the first endotherm appeared at lower temperature (30.6 ± 1.3 °C) and was broader, and the melting endotherm of KET was no longer observed. Upon heating at 50–300 °C/min TET showed a sharp melting event at 38 °C (data not shown). Hence, the eutectic and either KET or TET were present in the sample, but at a standard heating rate of 10 °C/min dissolution of KET and/or TET crystals dispersed in the liquid phase can take place resulting in a broad and flat endotherm.

After rapid cooling, TET formed a supercooled liquid with a glass transition at −53.6 ± 0.1 °C (Figure 6c,d). Upon second heating, supercooled TET showed a crystallization exotherm (onset: −5.2 ± 0.3 °C, ΔH = 102.8 ± 2.7 J/g) followed by a melting endotherm (onset: 37.8 ± 0.1 °C, ΔH = 116.4 ± 0.7 J/g). Since the value of crystallization ΔH is approximately 88% of melting ΔH, the crystallization occurs mainly during re-heating, but it is possible that the sample partially crystallized during the cooling step. The melting of TET in the second cycle occurred at a lower temperature compared with that from the first heating cycle (onsets 37.8 ± 0.2 °C and 42.2 ± 0.1 °C, respectively). Moreover, the PXRD pattern of QC TET that was heated past the crystallization in the DSC was different than that of the starting material (Figure 3b), indicating a different polymorphic form. Two polymorphs, II and I, have been reported for TET with melting points at 37 °C and 42 °C, respectively [37]. Hence, the starting material is most likely polymorph I (TET I), whereas the product obtained by crystallization during the second heating (heating of supercooled TET) is form II (TET II). The sample containing 5 mol% of KET formed a supercooled liquid and upon heating only 15% of TET crystallized into TET II, the rest of TET remained in the supercooled state. The TET crystallization peak for this sample appeared at higher temperatures (onset: 9.3 ± 4.4 °C) and was broader compared with that of 100% TET. The enthalpy of crystallization was the same as the enthalpy of melting, hence the crystallization process took place during heating. The QC sample containing 10 mol% of KET was disordered by PXRD, but a small quantity of crystalline TET (less than 1%) was detected by DSC. All other QC KET-TET mixtures (15–95 mol% of KET) exhibited only one phase change, a Tg, indicating mixing TET and KET at molecular level. The Tg values were dependent on the composition and they increased corresponding to an increasing quantity of KET, but in the KET-rich samples (90 and 95 mol% of KET) the Tgs reached a maximum and this temperature was the same as that of 100% KET. The Tgs of the QC mixtures samples containing at least 20 mol% of KET were higher than those calculated from Fox equation (Figure 6d), thereby implying important deviations from the behavior of an ideal system due to interactions between KET and TET.

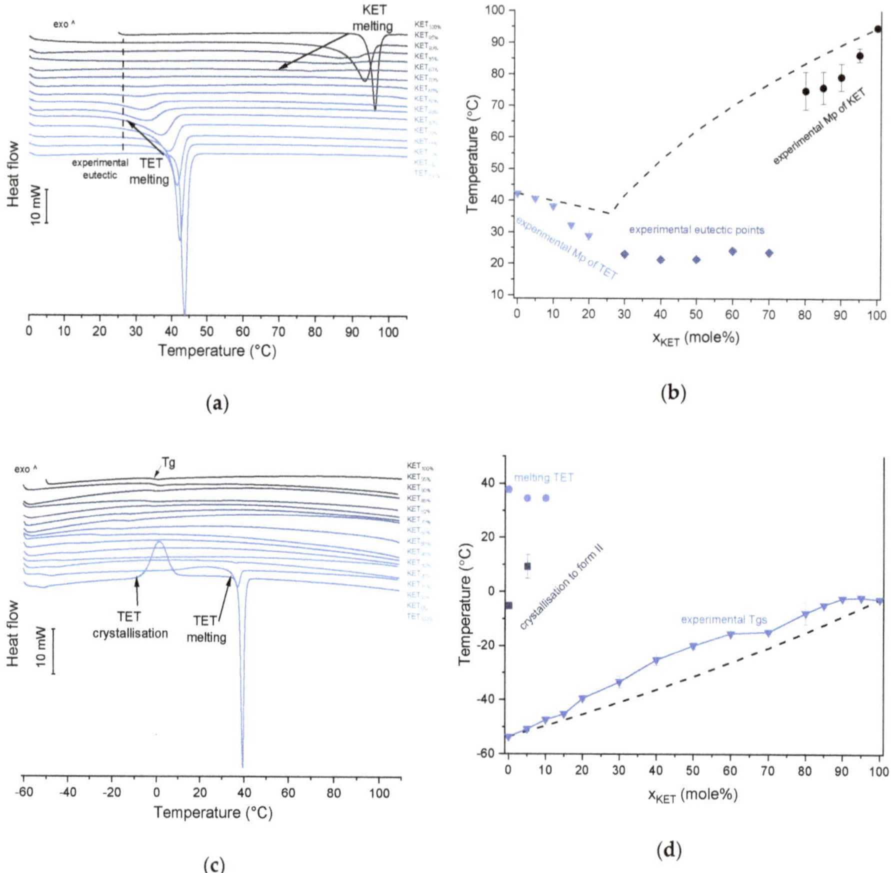

Figure 6. Thermal analysis of KET-TET systems: (**a**) DSC thermograms from first heating; the broken line indicates the position of the eutectic peak while the arrows show the position of the melting peaks, (**b**) phase diagram based on thermal analysis of first DSC heating; the broken lines show the theoretical liquidus curves calculated using Equation (1), (**c**) DSC thermograms from second heating (the samples were first heated 110 °C at 10 °C/min, quench cooled at a nominal cooling rate of 300 °C/min and reheated at 10 °C/min) (**d**) phase diagram based on thermal analysis of second DSC heating; the broken line shows the theoretical Tg values calculated using Equation (2). For plots (a) and (c) the subscript indicates the content of the named component in mole%.

The infrared spectrum of TET starting material (TET I) showed a single sharp band at 3371 cm^{-1} ascribed to the N–H stretching of the secondary aromatic amino group (Figure 5b), which shifted towards a higher wavenumber in TET II (3387 cm^{-1}). In the spectrum of TET I the stretching of C=O and C–O ester bonds produced bands at 1684 cm^{-1} and 1279 cm^{-1}, respectively, [35,38] and they shifted to lower wavenumbers (1681 cm^{-1} and 1261 cm^{-1}, respectively) in TET II. The presence of TET II was also observed in the spectrum of the mixture containing 5 mol% of KET. In the sample comprising 10 mol% of KET broadening and a dramatic decrease in the intensity of the N–H stretching band (3375 cm^{-1}) was observed due to TET molecules being able to interact with KET. The O–H stretching band was detected in the sample containing 90 mol% of KET, and it was difficult to observe it in the samples containing a smaller amount of KET. The position of the ester stretching C–O band was depended on the solid-state properties of the sample: it shifted from 1279 cm^{-1} (TET I) to 1261 cm^{-1} in

TET II, and in the QC KET-TET samples it was present at 1267 cm^{-1}. A decreased intensity of the ester C=O stretching band was observed in the QC samples. In the sample containing 10 mol% of KET it was localized at 1688 cm^{-1} and it shifted towards a higher wavenumber corresponding to an increasing KET concentration and overlapped with the carboxyl C=O stretching band. The latter band shifted towards a lower wavenumber and its intensity decreased corresponding to a decreasing KET concentration. The shoulder of the carboxyl C=O stretching band localized at 1737 cm^{-1} (100% KET) also shifted towards lower wavenumbers and was no longer observed in the sample containing 60 mol% of KET, similar to the KET-PRO system. The IR results indicate that TET interact with KET by strong hydrogen bonding, and the carboxyl group of KET is involved in the interactions. Interactions between KET and TET are sufficiently strong to break the crystal lattice of TET even at KET concentrations as low as 10 mol%. However, in contrast to KET-PRO, in the KET-TET samples carboxylate anion was not detected.

3.4. Ketoprofen-Benzocaine (KET-BEN) Systems

BEN exhibits polymorphism and three polymorphic forms have been described to date. Form I, formerly known as form β, is a monoclinic P 2_1/c polymorph (Z = 4) [39,40]. Form II, formerly known as α, is orthorhombic with space group $P2_12_12_1$ (Z = 4) [40,41]. Form III is another monoclinic $P2_1$ polymorph (Z = 8) [40,42]. Polymorph II (referred to as Mod I^0 by Schmidt, [43]) is thermodynamically stable under ambient conditions, and is present in commercial products. The starting material used in our study was form II. In the first heating cycle BEN showed a single melting endotherm (onset: 90.0 ± 0.1 °C). The KET-BEN mixtures formed a eutectic phase (Figure 7a). Both, KET and BEN melted at lower temperatures and had broader melting endotherms corresponding to their decreasing content in the mixture. Experimental Mps of KET and BEN showed an agreement with values calculated using the Schröder van Laar equation. All binary KET-BEN mixtures showed eutectic peaks with an onset at approximately 62–63 °C. The shape of the eutectic peak depended on the composition: the peaks were broader in the KET-rich samples than in the BEN-rich samples. It was difficult to discern individual peaks in the mixtures containing 30–70 mol% of KET. Thus, to construct the Tammann plot (ΔH of the eutectic peak as a function composition), only the mixtures containing 5–20 and 80–95 mol% of KET were taken. The eutectic composition determined using the Tammann plot was 34.8 of mol% of KET (Figure 7b). This composition contains less KET that the theoretical eutectic composition determined by Equation (1), which is 47 mol% KET. Assuming that this eutectic phase was a simple mechanical mix, the enthalpy of mixing (ΔH_{mix}) was calculated using Equation (4) [23]:

$$\Delta H_{mix} = (\Delta H_{fus})_{exp} - \left(x_{\mathrm{KET}} \cdot \Delta H_{fus\mathrm{KET}} + x_{\mathrm{BEN}} \cdot \Delta H_{fus\mathrm{BEN}}\right) \tag{4}$$

where $(\Delta H_{fus})_{exp}$ is the heat of fusion of the eutectic peak at the eutectic composition obtained from the Tammann plot, x is the mole fraction of the component (KET or BEN) and ΔH_{fus} is the heat of fusion of the pure constituent (KET or BEN). The value of ΔH_{mix} for the KET-BEN system was negative, −2.33 kJ/mol, indicating the presence of weak intermolecular forces leading to cluster formation [44]. The KET-BEN physical mixtures showed peaks characteristic of starting materials in X-ray diffractograms (Figure 3c).

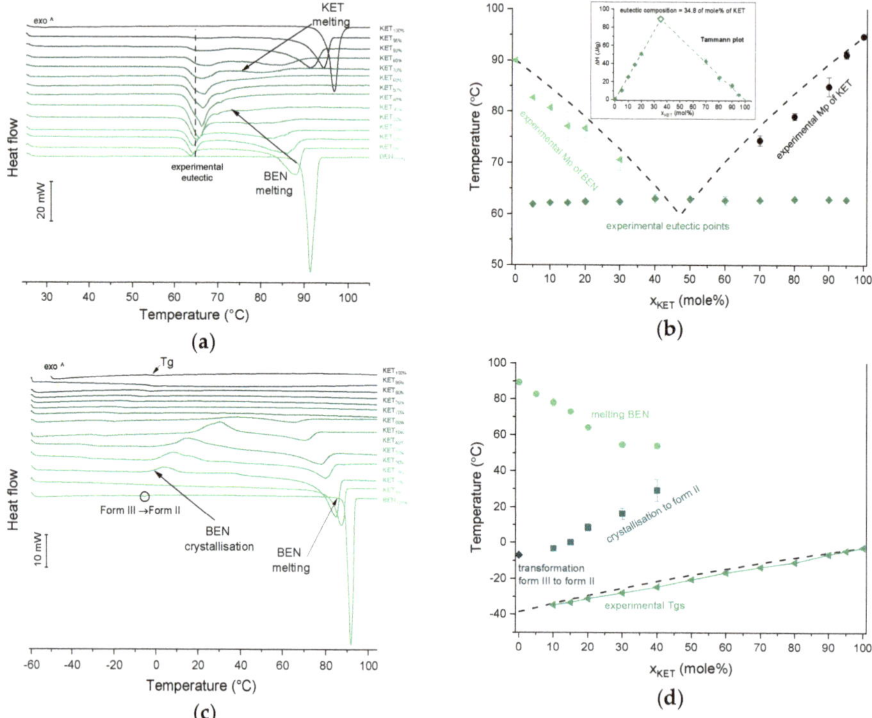

Figure 7. Thermal analysis of KET-BEN systems: (**a**) DSC thermograms from first heating; the broken line indicates the position of the eutectic peak while the arrows show the position of the melting peaks, (**b**) phase diagram based on thermal analysis of first DSC heating; the broken lines show the theoretical liquidus curves calculated using Equation (1). The Tammann plot is shown in the inset; the broken lines present the best linear fits to the experimental data, (**c**) DSC thermograms from second heating (the samples were first heated 110 °C at 10 °C/min, quench cooled at a nominal cooling rate of 300 °C/min and reheated at 10 °C/min) (**d**) phase diagram based on thermal analysis of second DSC heating; the broken line shows the theoretical Tg values calculated using Equation (2). For plots (a) and (c) the subscript indicates the content of the named component in mole%.

In the second heating cycle BEN melting endotherm was observed at the same temperature as in the first heating, but it was preceded by a small endotherm at −7.0 ± 0.1 °C (ΔH = 1.0 ± 0.2 J/g) (Figure 7c,d). Interestingly, Gana et al., [45] observed a similar endothermic event, which was attributed to a phase transition of form III into form II. Forms II and III are enantiotropic under ambient conditions. Form III is stable at low temperature and form II is stable at higher temperature and melts eventually. Form III becomes more stable with increasing pressure at ambient temperature [45]. Form II was detected in the QC BEN by PXRD (Figure 3c). After a rapid cooling step of the KET-BEN melt, a supercooled liquid was formed, except for the mixture containing 5 mol% of KET. A Tg was detectable down to the KET content of 10 mol%. A very strong linear relationship was observed between the composition (mol% of either BEN or KET) and the Tg (R^2 = 0.999). It was not achievable to obtain BEN 100% in the supercooled state, because it crystallized during cooling despite the very fast cooling rate employed. Hence, to determine the calculated Tgs using the Fox equation, the Tg of BEN 100% (−38.5 °C) was determined by extrapolation. The experimental Tgs showed a good agreement with the calculated Tgs. This implies the behavior of an ideal mixture, in which the tendency of two kinds of molecules (BEN and KET) to transfer from the glassy state to the supercooled liquid state is unchanged.

The presence of only one Tg in the binary mixtures containing at least 10 mol% of KET indicates that KET and BEN are miscible. The amount of crystalline of BEN decreased from 86.0 ± 0.7% (5 mol% of KET) to 15.2 ± 7.8% (40 mol% of KET), and crystalline BEN was no longer observed in samples containing at least 50 mol% of KET. QC mixtures that contained at least 50 mol% of KET showed a halo-pattern characteristic of disordered materials (Figure 3c). Hence, the increasing concentration of KET increased the amount of BEN entrapped in the supercooled state. In the sample containing 5 mol% of KET, crystallization of BEN took place during cooling. The crystallization of BEN was viewed during the second heating for mixtures containing 40–90 mol% of KET. The onset temperature of the crystallization peak increased corresponding to a decreasing BEN concentration in the sample. The percentage of BEN that crystallized during heating increased from 23% (10 mol% of KET) to approximately 100% in mixtures containing 20–40 mol% of KET. X-ray diffractograms confirmed that it is BEN that crystallizes, because only BEN peaks were observed in the X-ray diffractograms of heated, previously QC KET-BEN systems (KET ≤ 40 mol%) (Figure 3c). Some samples, such as that containing 10 mol% of KET, showed a multistep crystallization process, which could be the interfacial crystallization (first part of the peak) that was followed by bulk crystallization (second part of the peak). PXRD patterns of this QC sample that was heated in the second heating cycle to either 20 °C or 50 °C were the same and showed peaks characteristic for BEN (data not shown). The BEN crystallization peak was followed by the melting peak with an onset temperature decreasing corresponding to a decreasing BEN content, from 78.1 ± 2.3 °C to 54.0 ± 1.6 °C for KET 10 and 40 mol%, respectively. ΔH values of the BEN melting peak in the second heating cycle were markedly higher than those observed in the first heating cycle for the mixtures containing 5–30 mol% of KET.

The interactions between BEN and KET were further examined by ATR-FTIR (Figure 5c). The doublet consisting of two bands at 3421 and 3340 cm^{-1} can be ascribed to asymmetric and symmetric N–H stretching bands, respectively [35,40]. The scissoring band of NH_2 is present at 1595 cm^{-1}. The C=O and C–O bonds of aromatic esters produce bands at 1680 and 1273 cm^{-1}, respectively [35]. The latter may overlap with the C–N stretching band characteristic of aromatic amines in the range 1240–1366 cm^{-1} [35]. In the spectra of QC KET-BEN mixtures the positions of bands at 1273 cm^{-1} of the ester C-O stretching and at 1656 cm^{-1} of the stretching of ketone C=O group did not change markedly. In mixtures containing at least 50 mol% of KET the bands at 3421 and 3340 cm^{-1} of the asymmetric and symmetric N–H stretching bands disappeared, and new bands at 3472 and 3371 cm^{-1} were observed. The latter bands became noticeable in samples containing 20–30 mol% of KET. Interestingly, in the sample comprising KET 40 mol% both, the two former (3421 and 3340 cm^{-1}) and the two latter (3472 and 3371 cm^{-1}) bands were easily observed. The presence of those bands is in a good agreement with the crystallinity of BEN determined by PXRD and DSC. The packing of BEN molecules in crystals is stabilized by N–H···O hydrogen bonds [42]. Hence, the bands at 3421 and 3340 cm^{-1} correspond to N–H stretches of the NH_2 group in the crystal lattice that interact via H-bonding with the C=O moiety of the ester groups, and the increase in wavenumber to 3472 and 3371 cm^{-1} is probably due to the liberation of the NH_2 groups from the crystal lattice due to interactions with KET. This is consistent with the fact that the band of the ester C=O stretching of BEN (1680 cm^{-1}) was no longer detectable in the equimolar KET-BEN mixture. The peaks characteristic of the NH_2 group in the crystal lattice disappear in the samples that do not contain a detectable amount of crystalline BEN (at least 50 mol% of KET). The intensity of the bands characteristic of 'supercooled' NH_2 groups also agree well with the content of noncrystalline BEN in the sample. The band of the NH scissoring at 1595 cm^{-1} shifted towards higher wavenumbers (1600 cm^{-1}) in the QC samples containing 50–80 mol% of KET, the band shifted to 1597 cm^{-1} for the sample containing 90 mol% of KET because it was dominated by the C–C stretching band of KET at 1596 cm^{-1} [34]. The intensity of the broad O–H stretching band decreased corresponding to a decreasing KET concentration. Interestingly, this band was observed in the sample containing 60 mol% of KET, whereas in the PRO and TET mixtures it was only observed at KET 90 mol%. Also, both shoulders of the C=O stretching group band of the carboxyl acid monomer of KET (1704 and 1737 cm^{-1}) moved to lower wavenumber corresponding to a decreasing KET concentration

(1693 and 1727 cm^{-1} for the equimolar KET-BEN mixture). It can be concluded that KET interacts BEN via H-bonding, as suggested by the negative value of ΔH_{mix}, thereby decreasing the interactions between BEN molecules in the crystal lattice and leading to entrapment of BEN in the supercooled state.

3.5. DFT Studies

Molecular modelling studies were carried out to determine the likelihood and mode of intermolecular interactions in the binary systems. The global reactivity parameters are listed in Table 2. For LAs, the higher HOMO energy was calculated for TET, therefore this molecule should be the best electron donor, while the lowest LUMO energy was determined for BEN and PRO, implying the best electron acceptor properties of these two substances [46]. The values of I and A also suggested the same characteristics. Electronegativity values indicate if a molecule is a Lewis acid (large χ) or a base (low χ), therefore it can be said that KET will act as a Lewis acid (also consistent with the high value of ω) [47]. The molecule that is deemed as most reactive, based on the chemical hardness is TET followed closely by PRO. The global reactivity parameters, however, do not show a similar trend as that observed from the experimental studies, where PRO interacted with KET most strongly, followed by TET and BEN. It is because the calculated values do not include any effects of ionization (proton transfer), that possibly occur between KET and PRO.

Table 2. Dipole moments and global reactivity parameters of KET, PRO, TET and BEN. ΔE—HOMO/LUMO energy gap, I—electron affinity, A—ionization potential, χ—electronegativity, η—hardness and ω—electrophilicity index.

	Dipole moment	HOMO (eV)	LUMO (eV)	ΔE (eV)	I	A	X	η	ω
KET	2.4855	−7.048	−2.150	4.898	7.048	2.150	4.599	2.449	4.318
PRO	3.2811	−5.905	−1.143	4.762	5.905	1.143	3.524	2.381	2.608
TET	4.5044	−5.769	−1.034	4.735	5.769	1.034	3.4015	2.3675	2.444
BEN	3.4108	−6.068	−1.143	4.925	6.068	1.143	3.6055	2.4625	2.640

The molecular electrostatic potential (ESP) calculations can be vital in predicting intermolecular interactions [48]. Mapping the ESP on molecular vdW surface of KET and the studied LAs (Figure 8), clearly showed that the most likely intermolecular interactions between KET and each of the LAs will be between the –OH moiety of the carboxylic group of KET (ESP maximum of 48.5 kcal/mol) and the carbonyl of the ester moiety of the LA (ESP minimum of −38.13, −40.04 and −39.25 kcal/mol for PRO, TET and BEN, respectively). This is consistent with the results of the infrared analysis presented above, evidencing H-bond formation in the binary systems.

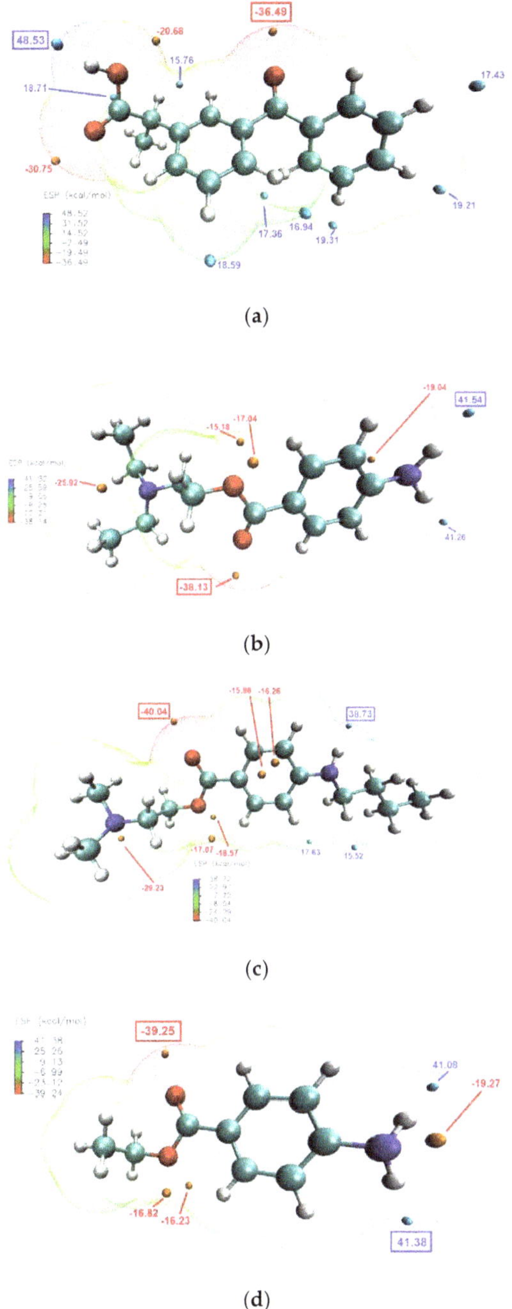

Figure 8. Electrostatic potential of: (**a**) KET, (**b**) PRO, (**c**) TET and (**d**) BEN mapped on the 0.001 au contours of the electron density of the molecules optimized by DFT. The negative regions are indicated in blue, while the positive regions are shown in red. Surface local minima and maxima (only values > 15 kcal/mol) are represented as orange and cyan points, respectively. The global maximum and minimum for every molecule are shown in boxes in large font. The unit is kcal/mol.

3.6. General Discussion

Multicomponent mixtures were successfully produced by the mechanosynthetic process, which involved mixing of an acid (KET) with a base (LA) in a range of molar ratios followed by heating. This process has important advantages. It does not necessitate the use of solvent, hence there is no need for product drying and determining the residual solvent content. Another advantage is the high product yield and purity because the process does not require product isolation and/or purification. Thus, the mechanosynthesis eliminates the drawbacks of commonly used metathesis reactions. The mechanosynthetic process described in this paper is applicable for NSAID-LA combinations and other DEMs/ILs. If the chosen combination contains substances that decompose upon melting (such as acetylsalicylic acid or diclofenac), solvent assisted grinding such as that described previously [20] for ethambutol ILs/salts or aspirin cocrystals [49] could be an alternative.

The heat treatment of a binary KET-LA physical mixtures resulted in the formation of eutectic phase. The thermograms of the binary KET-TET and KET-PRO mixtures contained broad endotherms with low intensity, except of mixtures containing very large excess of one of the components. It may be due to the fact that a part of dispersed crystals is surrounded by the liquid phase generated by the melting of the eutectic. This can induce the accelerated melting and/or dissolution of the component in excess (either KET or LA). Kataoka et al., [14] observed a similar phenomenon for ibuprofen-lidocaine binary mixture. The formation of eutectic mixtures has been observed for another NSAID–LA pair: indomethacin-lidocaine [18]. The eutectic peak in the KET-LA systems studied very often overlapped with the subsequent melting events. KET-PRO and KET-TET, except of LA-rich mixtures, showed important deviations from the behavior of an ideal system due to interactions between the KET and LA molecules, hence it was impossible to determine the eutectic composition. The behavior of the KET-BEN system, on the other hand, agreed with the theoretical predictions. The eutectic peaks were better separated from the succeeding thermal events compared with KET-TET and KET-PRO, and it was possible to determine the KET-BEN eutectic composition using the Tammann plot.

In the phase diagrams from the second heating, all KET-PRO samples were characterized by a single-phase transition from glass to the supercooled state. Similarly, only one Tg was observed in the samples containing 5–85 mol% of TET and 5–50 mol% of BEN. The presence of only one Tg confirms that KET and LAs are mixed at molecular level. KET-BEN and LA-rich KET-TET and KET-PRO mixtures showed a good agreement with theoretical predictions of Tg by Fox equation, however in the other KET-TET and KET-PRO combinations a considerable increase in Tg compared with predicted value was observed. This may imply the existence of strong KET-TET and KET-PRO interactions in the glassy state. During the second heating in the mixture containing 95 mol% of BEN only one melting was observed, whereas in the samples containing 60–90% of BEN and 95 mol% of TET three phase transitions were observed: a Tg, followed by LA crystallization and LA melting. The crystallization was always to the pure LA (either TET or BEN). Active pharmaceutical ingredients can be categorized into three different classes based on their crystallization tendency [16]. Despite the similarity of their structures, the tested LAs show a wide variation in crystallization tendency. BEN has been categorized as class I molecule, because it crystallizes during cooling from the undercooled melt state prior to the Tg event [16]. Hence, it has high crystallization tendency and low glass-forming ability. On the other hand, PRO and KET are class III molecules, with low crystallization tendency and high glass-forming ability for which no crystallization occurs upon either cooling to below Tg or upon subsequent reheating up to the melting point [16]. Baird et al. [16] did not investigate TET, but based on our results it can be categorized as a class II molecule, for which no crystallization is observed upon cooling from the undercooled melt state to below Tg, however, crystallization is observed during reheating above Tg. Although BEN has a very low glass-forming ability, after mixing with KET it was possible to capture some BEN molecules in the supercooled state. Interactions between KET and TET are sufficiently strong to break the crystal lattice of TET even at KET concentrations as low as 10 mol%.

The preparation of protic ILs typically demands a sufficient pK_a difference between the acid and the base that could lead to an effective proton transfer and formation of an ion pair [6]. The recommended difference of ΔpK_a of 10 is not possible to achieve for most APIs [6], including NSAIDs and LAs. The pK_a is determined in a diluted aqueous solution, hence its application for tested multicomponent systems in the pure form that are not dissolved/dispersed in water, is limited. The ionization depends not only on the pK_a, but also on structural features of acid and base [6,8,17]. The DSC and IR results show that interactions between KET and BEN are markedly weaker than those between KET and either TET or PRO. This is consistent with the fact that BEN is weaker base than either of TET or PRO. The interactions between PRO and KET seem to be stronger than those between TET and KET, because there are indications of the presence of a small level of carboxylate anions in the former mixture. Moreover, the positive deviations from the theoretical Tgs that are larger in KET-PRO system (a difference of 16 °C for the equimolar mixture) than in KET-TET system (a difference of 11 °C for mixtures containing 40–60 mol% of KET) imply stronger electrostatic interactions in the former. In systems composed of ciprofloxacin and Eudragit L100, particularly large positive *Tg* deviations (45–48 °C) from theoretical predictions were observed due to ionic interactions [21]. The low degree of proton transfer in the KET-LA systems tested is consistent with the previous observations for similar systems such as ibuprofen-lidocaine [2]. When a tertiary amine is used as the base (PRO and TET in our study) the proton transfer may be severely restricted because of the lack of a satisfactory hydrogen bonding solvation environment for the anionic species formed [2]. In conclusion, all KET-LA systems studies in this work are DEMs, i.e., mixtures of hydrogen bond donors and acceptors with intermolecular interactions via hydrogen bonding.

Although the salt preparation of the APIs with poor solubility in water is one of the most effective and developable approach to improve their solubility and dissolution rate, new salts of the API may be recognized as new chemical entities by the FDA and other healthcare authorities. Therefore, the absence of ion pair formation in the KET-LA systems could be beneficial from the regulatory point of view, as they should not be considered as new chemical entities.

4. Conclusions

A mechanosynthetic process that involved mixing and co-melting of a free acid and a free base proved to be applicable for producing KET-LA DEMs. The binary KET-LA powder mixtures formed eutectic phases. After a rapid cooling the formation of a supercooled liquid, with a *Tg* that was highly dependent on the composition, was observed for all mixtures with exception of that containing 95% mol% of BEN. Crystallization of LA was observed for QC samples comprising BEN-rich mixtures (BEN $\geq$ 60 mol%) and the sample containing TET 95 mol%. In the presence of KET the high crystallization tendency of LAs of TET and BEN can be decreased, and their glass-forming ability increased. The KET-PRO DEMs in the entire composition range did not crystallize either during the cooling step or during the second heating cycle. Experimental Mps and Tgs of KET-BEN mixtures follow the theoretical Schöder van Laar and Fox prediction, respectively. On the other hand, for PRO and TET-based systems, high deviations from the theoretical Schröder van Laar and Fox predictions were observed, indicating the formation of strong H-bonded complexes between these LAs and KET. A small quantity of carboxylate anions was present in the KET-PRO samples, but the proton transfer is severely restricted due to the lack of a satisfactory hydrogen bonding solvation environment necessary for anion stabilization. The strength of interactions with KET can be ranked in the following order: PRO > TET > BEN. Therefore, it can be concluded that KET-LA mixtures do not form ILs, but DEMs.

Author Contributions: Conceptualization, A.U., K.B., J.Z., M.S. and L.T.; methodology, A.U., K.B., J.Z. and M.S.; formal analysis, A.U., K.B., J.Z. M.S. and L.T.; investigation, A.U., K.B., J.Z. and M.S.; writing-original draft preparation, A.U.; writing-review and editing, L.T.; visualization, A.U. and L.T.; supervision, L.T.; project administration, L.T.; funding acquisition, L.T. All authors have read and agreed to the published version of the manuscript.

Funding: This research was funded by Science Foundation Ireland, grant number 15/CDA/3602, and support was provided by the Synthesis and Solid State Pharmaceutical Centre (SSPC), financed by a research grant from Science Foundation Ireland (SFI) and co-funded under the European Regional Development Fund (grant number 12/RC/2275).

Conflicts of Interest: The authors declare no conflict of interest. The funders had no role in the design of the study; in the collection, analyses, or interpretation of data; in the writing of the manuscript, or in the decision to publish the results.

References

1. Ferraz, R.; Branco, L.C.; Prudêncio, C.; Noronha, J.P.; Petrovski, Ž. Ionic liquids as active pharmaceutical ingredients. *ChemMedChem* **2011**, *6*, 975–985. [CrossRef]
2. Wang, H.; Gurau, G.; Shamshina, J.; Cojocaru, O.A.; Janikowski, J.; Macfarlane, D.R.; Davis, J.H.; Rogers, R.D. Simultaneous membrane transport of two active pharmaceutical ingredients by charge assisted hydrogen bond complex formation. *Chem. Sci.* **2014**, *5*, 3449–3456. [CrossRef]
3. Hough, W.L.; Smiglak, M.; Rodríguez, H.; Swatloski, R.P.; Spear, S.K.; Daly, D.T.; Pernak, J.; Grisel, J.E.; Carliss, R.D.; Soutullo, M.D.; et al. The third evolution of ionic liquids: active pharmaceutical ingredients. *New J. Chem.* **2007**, *31*, 1429. [CrossRef]
4. Hough, W.L.; Rogers, R.D. Ionic liquids then and now: From solvents to materials to active pharmaceutical ingredients. *Bull. Chem. Soc. Jpn.* **2007**, *80*, 2262–2269. [CrossRef]
5. Bica, K.; Rijksen, C.; Nieuwenhuyzen, M.; Rogers, R.D. In search of pure liquid salt forms of aspirin: ionic liquid approaches with acetylsalicylic acid and salicylic acid. *Phys. Chem. Chem. Phys.* **2010**, *12*, 2011–2017. [CrossRef]
6. Balk, A.; Holzgrabe, U.; Meinel, L. Pro et contra' ionic liquid drugs—Challenges and opportunities for pharmaceutical translation. *Eur. J. Pharm. Biopharm.* **2015**, *94*, 291–304. [CrossRef]
7. Balk, A.; Wiest, J.; Widmer, T.; Galli, B.; Holzgrabe, U.; Meinel, L. Transformation of acidic poorly water soluble drugs into ionic liquids. *Eur. J. Pharm. Biopharm.* **2015**, *94*, 73–82. [CrossRef]
8. Stoimenovski, J.; MacFarlane, D.R.; Bica, K.; Rogers, R.D. Crystalline vs. ionic liquid salt forms of active pharmaceutical ingredients: A position paper. *Pharm. Res.* **2010**, *27*, 521–526. [CrossRef]
9. Wojnarowska, Z.; Smolka, W.; Zotova, J.; Knapik-Kowalczuk, J.; Sherif, A.; Tajber, L.; Paluch, M. The effect of electrostatic interactions on the formation of pharmaceutical eutectics. *Phys. Chem. Chem. Phys.* **2018**, *20*, 27361–27367. [CrossRef]
10. Park, H.J.; Prausnitz, M.R. Lidocaine-ibuprofen ionic liquid for dermal anesthesia. *AIChE J.* **2015**, *61*, 2732–2738. [CrossRef]
11. Wang, C.; Chopade, S.A.; Guo, Y.; Early, J.T.; Tang, B.; Wang, E.; Hillmyer, M.A.; Lodge, T.P.; Sun, C.C. Preparation, Characterization, and Formulation Development of Drug-Drug Protic Ionic Liquids of Diphenhydramine with Ibuprofen and Naproxen. *Mol. Pharm.* **2018**, *15*, 4190–4201. [CrossRef]
12. Kong, D.X.; Li, X.J.; Zhang, H.Y. Where is the hope for drug discovery? Let history tell the future. *Drug Discov. Today* **2009**, *14*, 115–119. [CrossRef]
13. Beaulieu, P. Non-opioid strategies for acute pain management. *Can. J. Anesth. Can. D'anesthésie* **2009**, *54*, 481–485. [CrossRef]
14. Kataoka, H.; Sakaki, Y.; Komatsu, K.; Shimada, Y.; Goto, S. Melting Process of the Peritectic Mixture of Lidocaine and Ibuprofen Interpreted by Site Percolation Theory Model. *J. Pharm. Sci.* **2017**, *106*, 3016–3021. [CrossRef]
15. Ribeiro, R.; Pinto, P.C.A.G.; Azevedo, A.M.O.; Bica, K.; Ressmann, A.K.; Reis, S.; Saraiva, M.L.M.F.S. Automated evaluation of protein binding affinity of anti-inflammatory choline based ionic liquids. *Talanta* **2016**, *150*, 20–26. [CrossRef]
16. Baird, J.A.; Van Eerdenbrugh, B.; Taylor, L.S. A classification system to assess the crystallization tendency of organic molecules from undercooled melts. *J. Pharm. Sci.* **2010**, *99*, 3787–3806. [CrossRef]
17. Moreira, D.N.; Fresno, N.; Pérez-Fernández, R.; Frizzo, C.P.; Goya, P.; Marco, C.; Martins, M.A.P.; Elguero, J. Brønsted acid-base pairs of drugs as dual ionic liquids: NMR ionicity studies. *Tetrahedron* **2015**, *71*, 676–685. [CrossRef]
18. Shimada, Y.; Goto, S.; Uchiro, H.; Hirabayashi, H.; Yamaguchi, K.; Hirota, K.; Terada, H. Features of heat-induced amorphous complex between indomethacin and lidocaine. *Colloids Surf. B Biointerfaces* **2013**, *102*, 590–596. [CrossRef]

19. Fiandaca, M.; Dalwadi, G.; Wigent, R.; Gupta, P. Ionic liquid formation with deep eutectic forces at an atypical ratio (2:1) of naproxen to lidocaine in the solid-state, thermal characterization and FTIR investigation. *Int. J. Pharm.* **2020**, *575*, 118946. [CrossRef]
20. Cherukuvada, S.; Nangia, A. Salts and ionic liquid of the antituberculosis drug S,S-ethambutol. *Cryst. Growth Des.* **2013**, *13*, 1752–1760. [CrossRef]
21. Mesallati, H.; Umerska, A.; Paluch, K.J.; Tajber, L. Amorphous Polymeric Drug Salts as Ionic Solid Dispersion Forms of Ciprofloxacin. *Mol. Pharm.* **2017**, *14*, 2209–2223. [CrossRef]
22. Umerska, A.; Paluch, K.J.; Santos-Martinez, M.J.; Corrigan, O.I.; Medina, C.; Tajber, L. Freeze drying of polyelectrolyte complex nanoparticles: Effect of nanoparticle composition and cryoprotectant selection. *Int. J. Pharm.* **2018**, *552*, 27–38. [CrossRef]
23. Meltzer, V.; Pincu, E. Thermodynamic study of binary mixture of citric acid and tartaric acid. *Cent. Eur. J. Chem.* **2012**, *10*, 1584–1589. [CrossRef]
24. Pinal, R. Entropy of mixing and the glass transition of amorphous mixtures. *Entropy* **2008**, *10*, 207–223. [CrossRef]
25. Mugheirbi, N.A.; Paluch, K.J.; Tajber, L. Heat induced evaporative antisolvent nanoprecipitation (HIEAN) of itraconazole. *Int. J. Pharm.* **2014**, *471*, 400–411. [CrossRef]
26. Frisch, M.; Trucks, G.; Schlegel, H.; Scuseria, G. *Gaussian 03, Revision C. 02*; Gaussian, Inc.: Wallingford, CT, USA, 2004.
27. Koopmans, T. Über die Zuordnung von Wellenfunktionen und Eigenwerten zu den Einzelnen Elektronen Eines Atoms. *Physica* **1934**, *1*, 104–113. [CrossRef]
28. Klopman, G. Chemical Reactivity and the Concept of Charge- and Frontier-Controlled Reactions. *J. Am. Chem. Soc.* **1968**, *90*, 223–234. [CrossRef]
29. Mesallati, H.; Mugheirbi, N.A.; Tajber, L. Two Faces of Ciprofloxacin: Investigation of Proton Transfer in Solid State Transformations. *Cryst. Growth Des.* **2016**, *16*, 6574–6585. [CrossRef]
30. Lu, T.; Chen, F. Multiwfn: A multifunctional wavefunction analyzer. *J. Comput. Chem.* **2012**, *33*, 580–592. [CrossRef]
31. Lu, T.; Chen, F. Quantitative analysis of molecular surface based on improved Marching Tetrahedra algorithm. *J. Mol. Graph. Model.* **2012**, *38*, 314–323. [CrossRef]
32. Wishart, D.S. DrugBank: a comprehensive resource for in silico drug discovery and exploration. *Nucleic Acids Res.* **2006**, *34*, D668–D672. [CrossRef]
33. Champeau, M.; Thomassin, J.M.; Jérôme, C.; Tassaing, T. In situ investigation of supercritical CO_2 assisted impregnation of drugs into a polymer by high pressure FTIR micro-spectroscopy. *Analyst* **2015**, *140*, 869–879. [CrossRef]
34. Vueba, M.L.; Pina, M.E.; Veiga, F.; Sousa, J.J.; De Carvalho, L.A.E.B. Conformational study of ketoprofen by combined DFT calculations and Raman spectroscopy. *Int. J. Pharm.* **2006**, *307*, 56–65. [CrossRef]
35. Fuliaş, A.; Ledeţi, I.; Vlase, G.; Popoiu, C.; Hegheş, A.; Bilanin, M.; Vlase, T.; Gheorgheosu, D.; Craina, M.; Ardelean, S.; et al. Thermal behaviour of procaine and benzocaine Part II: Compatibility study with some pharmaceutical excipients used in solid dosage forms. *Chem. Cent. J.* **2013**, *7*, 1–10. [CrossRef]
36. Kashino, S.; Ikeda, M.; Haisa, M. The structure of procaine. *Acta Crystallogr. Sect. B Struct. Crystallogr. Cryst. Chem.* **1982**, *38*, 1868–1870. [CrossRef]
37. Schmidt, A.C. The role of molecular structure in the crystal polymorphism of local anesthetic drugs: Crystal polymorphism of local anesthetic drugs, part X. *Pharm. Res.* **2005**, *22*, 2121–2133. [CrossRef]
38. Shibata, A.; Ikawa, K.; Terada, H. Site of action of the local anesthetic tetracaine in a phosphatidylcholine bilayer with incorporated cardiolipin. *Biophys. J.* **1995**, *69*, 470–477. [CrossRef]
39. Lynch, D.E.; McClenaghan, I. Monoclinic form of ethyl 4-aminobenzoate (benzocaine). *Acta Crystallogr. Sect. E Struct. Rep. Online* **2002**, *58*, o708–o709. [CrossRef]
40. Paczkowska, M.; Wiergowska, G.; Miklaszewski, A.; Krause, A.; Mroczkowka, M.; Zalewski, P.; Cielecka-Piontek, J. The Analysis of the Physicochemical Properties of Benzocaine Polymorphs. *Molecules* **2018**, *23*, 1737. [CrossRef]
41. Chan, E.J.; Rae, A.D.; Welberry, T.R. On the polymorphism of benzocaine; A lowtemperature structural phase transition for form (II). *Acta Crystallogr. Sect. B Struct. Sci.* **2009**, *65*, 509–515. [CrossRef]

42. Sinha, B.K.; Pattabhi, V. Crystal structure of benzocaine-A local anaesthetic. *J. Chem. Sci.* **1987**, *98*, 229–234. [CrossRef]
43. Schmidt, A.C. Structural characteristics and crystal polymorphism of three local anaesthetic bases—Crystal polymorphism of local anaesthetic drugs: Part VII. *Int. J. Pharm.* **2005**, *298*, 186–197. [CrossRef]
44. Rastogi, R.P.; Singh, N.B.; Dwivedi, K.D. Solidification behaviour of addition compounds and eutectics of pure components and addition compounds. *Ber. Bunsenges. Phys. Chem.* **1981**, *85*, 85–91. [CrossRef]
45. Gana, I.; Barrio, M.; Do, B.; Tamarit, J.L.; Céolin, R.; Rietveld, I.B. Benzocaine polymorphism: Pressure-temperature phase diagram involving forms II and III. *Int. J. Pharm.* **2013**, *456*, 480–488. [CrossRef]
46. Gázquez, J.L.; Cedillo, A.; Vela, A. Electrodonating and electroaccepting powers. *J. Phys. Chem. A* **2007**, *111*, 1966–1970. [CrossRef]
47. Pearson, R.G. Absolute electronegativity and hardness correlated with molecular orbital theory. *Proc. Natl. Acad. Sci. USA* **1986**, *83*, 8440–8441. [CrossRef]
48. Murray, J.S.; Politzer, P. The electrostatic potential: An overview. *Wiley Interdiscip. Rev. Comput. Mol. Sci.* **2011**, *1*, 153–163. [CrossRef]
49. Darwish, S.; Zeglinski, J.; Krishna, G.R.; Shaikh, R.; Khraisheh, M.; Walker, G.M.; Croker, D.M. A New 1:1 Drug-Drug Cocrystal of Theophylline and Aspirin: Discovery, Characterization, and Construction of Ternary Phase Diagrams. *Cryst. Growth Des.* **2018**, *18*, 7526–7532. [CrossRef]

Article

Alendronic Acid as Ionic Liquid: New Perspective on Osteosarcoma

Sónia Teixeira [1,†], Miguel M. Santos [2,†], Maria H. Fernandes [1,3], João Costa-Rodrigues [1,4,5] and Luís C. Branco [2,*]

1 Faculdade de Medicina Dentária, U. Porto, Rua Dr. Manuel Pereira da Silva, 4200-393 Porto, Portugal; soniateixeira.88@gmail.com (S.T.); mhfernandes@fmd.up.pt (M.H.F.); joao.m.rodrigues@gmail.com (J.C.-R.)
2 LAQV-REQUIMTE, Departamento de Química, Faculdade de Ciências e Tecnologia, Universidade Nova de Lisboa, 2829-516 Caparica, Portugal; miguelmsantos@fct.unl.pt
3 UCIBIO-REQUIMTE, Departamento de Química, Faculdade de Ciências, Universidade do Porto, 4169-007 Porto, Portugal
4 ESS—Escola Superior de Saúde, Instituto Politécnico do Porto, R. Dr. António Bernardino de Almeida 400, 4200-072 Porto, Portugal
5 Instituto Politécnico de Viana do Castelo, Escola Superior de Saúde, Rua D. Moisés Alves Pinho 190, 4900-314 Viana do Castelo, Portugal
* Correspondence: l.branco@fct.unl.pt
† These authors contributed equally to the article.

Received: 31 December 2019; Accepted: 16 March 2020; Published: 24 March 2020

Abstract: Herein the quantitative synthesis of eight new mono- and dianionic Organic Salts and Ionic Liquids (OSILs) from alendronic acid (ALN) is reported by following two distinct sustainable and straightforward methodologies, according to the type of cation. The prepared ALN-OSILs were characterized by spectroscopic techniques and their solubility in water and biological fluids was determined. An evaluation of the toxicity towards human healthy cells and also human breast, lung and bone (osteosarcoma) cell lines was performed. Globally, it was observed that the monoanionic OSILs showed lower toxicity than the corresponding dianionic structures to all cell types. The highest cytotoxic effect was observed in OSILs containing a [C_2OHMIM] cation, in particular [C_2OHMIM][ALN]. The latter showed an improvement in IC_{50} values of ca. three orders of magnitude for the lung and bone cancer cell lines as well as fibroblasts in comparison with ALN. The development of OSILs with high cytotoxicity effect towards the tested cancer cell types, and containing an anti-resorbing molecule such as ALN may represent a promising strategy for the development of new pharmacological tools to be used in those pathological conditions.

Keywords: alendronic acid; Active Pharmaceutical Ingredients; API-OSILs; anticancer drugs; ionic liquids

1. Introduction

Alendronic acid is an aminobisphosphonate derivative that has shown efficacy in postmenopausal osteoporosis, malignant hypercalcemia and Paget's disease [1]. Alendronate localizes preferentially at active sites of bone resorption, which has been inhibited at doses that have no effect on bone mineralization [2].

Bisphosphonates bind at the bone mineral surface, where they potently inhibit osteoclast-mediated bone resorption and subsequently embed in the bone, being released only during subsequent resorption [3]. In contrast to other antiresorptive agents, bisphosphonates with the greatest binding affinity to bone (zoledronic acid > alendronate > ibandronate > risedronate) may persist in bone, and patients continue to be exposed to the pharmacologic effects of these drugs several years after discontinuation.

All bisphosphonates rapidly reduce bone resorption, which leads to decreased bone formation because resorption and formation are coupled. Within three to six months, equilibrium is reached at a lower rate of bone turnover [4].

It is described that alendronate in rats exhibited 200 to 1000 times more potency than etidronate and approximately 100 times more in comparison with clodronate or tiludronate [5]. It is recognized that the presence of amino group side-chain from alendronate chemical structure contributes to greater potency and specificity [3–5].

The introduction of specific functional groups in the BPs structure can lead to modifications in their physicochemical, biological, therapeutic and toxicological properties. In recent literature studies, alendronic acid is reported as an FDA drug already approved for the prevention and treatment of osteoporosis in men and women, either postmenopausal or glucocorticoid-induced [6]. However, BPs present low bioavailability when administered orally and frequently require parenteral administration, which is not the most convenient route in case of continuous treatment.

The combination of Active Pharmaceutical Ingredients (APIs) with biocompatible organic counter-ions has rendered over the last decade an attractive class of materials entitled API–Organic Salts and Ionic Liquids (API-OSILs) [7–10]. Ionic Liquids (ILs) are defined as salts with a melting point below 100 °C, which display peculiar properties such as negligible vapor pressure, high thermal and chemical stability and tunable physicochemical properties according to the cation-anion combinations. These novel API–OSILs can improve the drug performance in terms of stability, solubility, permeability, biological activity and delivery [10–22]. Recent achievements showed that the suitable combination between different families of pharmaceutical drugs such as NSAIDs (e.g., ibuprofen and naproxen [15,16]), β-lactam (e.g., ampicillin, amoxicillin, penicillin [16–20]) and fluoroquinolone (ciprofloxacin, norfloxacin [16,21]) antibiotics and bone antiresorptive agents (zoledronic acid [22]) rendered novel pharmaceutical drug formulations based on OSILs. Elimination of original polymorphism, improvement of permeability and solubility in water and biological fluids as well as increased therapeutic efficiency was observed for these innovative compounds.

Thus, considering the pharmaceutical importance of bisphosphonate drugs as antiresorptive and potential antitumoral agents, it is of paramount importance that novel formulations of such drugs, which render higher bioavailability and lower systemic toxicity than the latter are developed. Hence, in this paper, we report our latest results on our research line regarding bisphosphonate-based OSILs. In particular, we describe the synthesis of eight new Ionic Liquids and Organic Salts from alendronic acid (ALN–OSILs) as mono- and dianion by combination with one or two units of biocompatible cations, respectively. The desired compounds were prepared in quantitative yields by two distinct sustainable and straightforward methodologies, according to the type of cation. All prepared ALN–OSILs were characterized by spectroscopic techniques and their solubility in water and biological fluids was determined. Finally, evaluation of the toxicity towards human healthy cells and lung, breast and bone (osteosarcoma) cell lines was performed.

2. Synthesis and Characterization of ALN–OSILs

Figure 1 depicts the structures of ALN and the selected cations, which were combined in 1:1 or 1:2 stoichiometric ratios, thereby deprotonating one or two phosphonate groups, respectively.

The selection of protonated organic superbases (1,1,3,3-tetramethylguanidinium [TMGH] and 1,5-diazabicyclo(4.3.0)non-5-enium [DBNH]) and quaternary ammonium (cholinium [Ch] and 1-(2-hydroxyethyl)-3-methyl-1*H*-imidazol-3-ium [C_2OHMIM]) cations relied on the known information about their biocompatibility [15,23]. Despite their low toxicity, the combination of these cations with other APIs has rendered highly active API–OSILs in the past, in particular antimicrobial and anti-tumoral [19,20,22].

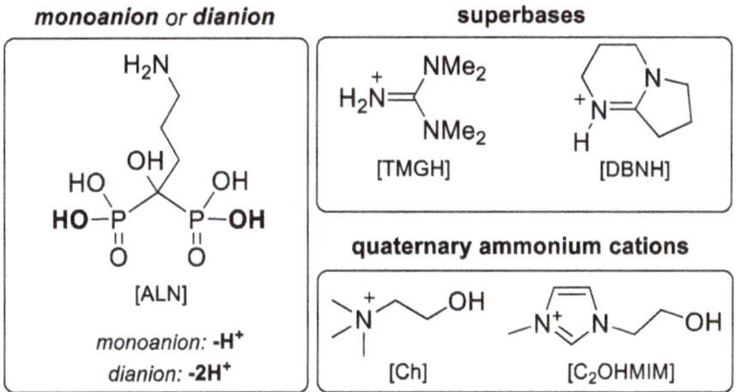

Figure 1. Structure of mono- and dianionic alendronate, protonated superbases and organic quaternary cations used to prepare the alendronic acid (ALN)–Organic Salts and Ionic Liquids (OSILs).

Scheme 1 resumes the synthetic methodologies employed in the synthesis of the API–OSILs.

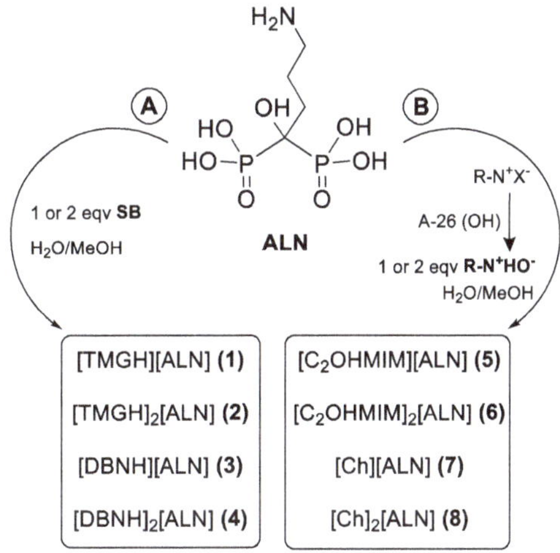

Scheme 1. Methodologies (**A**,**B**) for the preparation of ALN–OSILs.

The preparation of compounds **1–4** consisted of the addition of the diluted superbases to a dispersion of ALN (pathway A). These particular compounds, termed Superionic Liquids of APIs, were prepared according to a previous methodology reported by us for zoledronic acid [22] as well as other APIs [16]. While both monoanionic ALN–OSILs were obtained as solids, the two dianionic were isolated as pastes.

On the other hand, a combination of ALN with one and two equivalents of [Ch] and [C_2OHMIM] cations yielded four more ALN-based OSILs (compounds **5–8**) by following another methodology already described by our group [15,22]. In this case, quaternary ammonium hydroxide cations are previously prepared by the corresponding halide exchange with hydroxide exchange resins (e.g., Amberlyst A26-OH) in methanolic solution. The very basic solutions were then neutralized by addition to an aqueous solution of bisphosphonate yielding the desired salts in quantitative yields. From the four synthesized compounds, two are RTILs while the other two are solids (see below).

All products were characterized by NMR (^{1}H and ^{13}C) and FTIR spectroscopic techniques, as well as elemental analysis. The thermal properties were evaluated by DSC and the solubility in water and saline solution was determined for all compounds.

The NMR spectra of the ALN–OSILs were acquired in D_2O taking advantage of their high solubility in water (see below). In all cases, the ^{1}H NMR spectra showed that the cation/anion proportion is strictly 1.0:1.0 or 2.0:1.0, in agreement with the intended stoichiometry (see Figures S1–S16). In addition, only one set of signals was observed, meaning that the reactions were complete and only one product was formed. No comparison with alendronic acid is achievable because of its lack of solubility in the same solvents as the ALN–OSILs. In the ^{13}C NMR spectra, the resonance of the quaternary carbon atom of ALN appears at ca. 74 ppm, with no particular difference if only one or two neighboring phosphonate groups are deprotonated. Similarly, the ^{1}H NMR data is also irreflective of the ionic state of the bisphosphonates and also of the cations. In contrast, the collected FTIR spectra (Figures S17–S25) show pristine variations between the neutral bisphosphonates and the ALN–OSILs, as well as between mono- and dianions. The FTIR spectrum of ALN shows two characteristic regions, namely a weak and undefined structure at 2400–2000 cm^{-1} with maximum intensity at 2256 cm^{-1}, and also very intense multiple peaks at 1250–900 cm^{-1} (see Figure 2A). While the first one accounts for O–H stretches from the O=P–OH groups, the second region contains peaks assignable to the stretch of both P=O and P–OH bonds [24]. In the spectra of the synthesized ALN–OSILs (see Figure 2 for [TMGH]-based ALN–OSILs), the peaks in the first region become sharper and much weaker, and two other sets of peaks appear in the vicinity (at ca. 2400–2300 cm^{-1} and 2200–2100 cm^{-1}) which are more intense for the dianionic ALN–OSILs than for the monoanionic. This is in complete agreement with changes in the vibrational modes of the OH groups from the phosphonate moieties that are particularly enhanced when both groups are deprotonated. This corollary is also sustained by the changes observed in the second region of peaks, which in general displays two very intense broad peaks at ca. 1160 cm^{-1} and 1060 cm^{-1}, and one or two of medium intensity between 960 and 900 cm^{-1}.

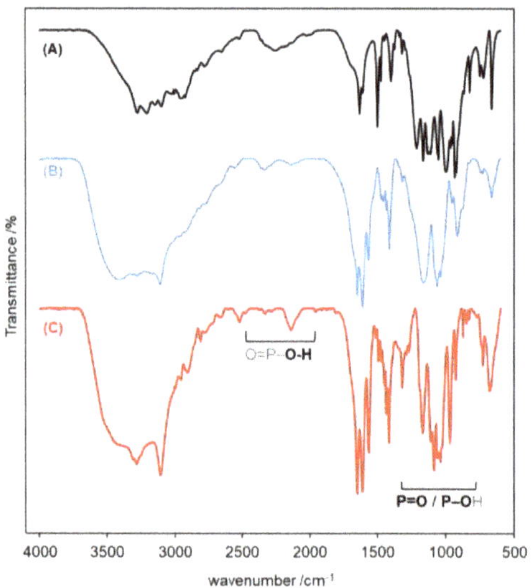

Figure 2. FTIR spectra of (**A**) ALN, (**B**) [TMGH][ALN] and (**C**) $[TMGH]_2[ALN]$.

3. Thermal Analysis of ALN–OSILs

All prepared OSILs from alendronic acid were studied by Differential Scanning Calorimetry (DSC) techniques (see Figures S26–S33). Table 1 contains the obtained data, namely melting, crystallization and glass transition temperatures, as well as the physical state of the analyzed compounds.

In general, the monoanionic ALN–OSILs are foam-like solids while the dianionic ones were obtained as thick colorless pastes at room temperature, thus being considered Room Temperature Ionic Liquids (RTILs). The exceptions to this rule are the compounds with the $[C_2OHMIM]$ cation, where the mono- and dianionic are, respectively, a paste and a foam. This inversion of the trend is probably related to specific interactions of ALN with the electron-rich imidazolium ring [25].

Table 1. Physical state, melting (T_m), cold crystallization (T_{cc}) and glass transition (T_g) temperatures of the prepared ALN–OSILs.

Salt	Physical State	T_m/°C	T_{cc}/°C *	T_g/°C
[TMGH][ALN]	White solid	162.7	107.1	-
[TMGH]$_2$[ALN]	Colorless paste	-	-	97.5
[DBNH][ALN]	White solid	130.3; 133.2	-	-
[DBNH]$_2$[ALN]	Colorless paste	-	-	45.7
[C$_2$OHMIM][ALN]	Colorless paste	-	-	64.5
[C$_2$OHMIM]$_2$[ALN]	White solid	153.0 (dec)	-	46.3
[Ch][ALN]	White solid	141.2	-	74.9
[Ch]$_2$[ALN]	Colorless paste	-	-	63.8

* Cold crystallization.

In comparison with sodium alendronate, which presents a melting temperature of 259.3 °C [26], all solid compounds melt at lower temperatures, more specifically between 130.3 and 162.7 °C, for [DBNH]- and [TMGH][ALN] OSILs, respectively. These melting temperatures are determined in the third cycle of the experiment, which consists of heating from −90 °C to 150–190 °C (depending on the compound) at 10 °C/min. The first two cycles typically consist of heating to 125 °C and isotherm for 15–20 min (for the complete removal of residual water) followed by cooling to −90 °C. From Cycle 3 onwards, consecutive cooling/heating cycles are performed at 10 or 20 °C/min. These cycles show glass transition temperatures (T_g) for the majority of the ALN–OSILs, showing that they become supercooled products, i.e., amorphous after the first melt. The exceptions to this behavior are [TMGH] and [DNBH][ALN], to which no glass transition nor crystallization (T_c) temperatures were observed in the experimental conditions. However, [TMGH][ALN] showed different behavior in comparison with the remaining ALN–OSILs in the first cycle, in particular a cold crystallization (T_{cc}) peak at 107.1 °C (Figure 3A).

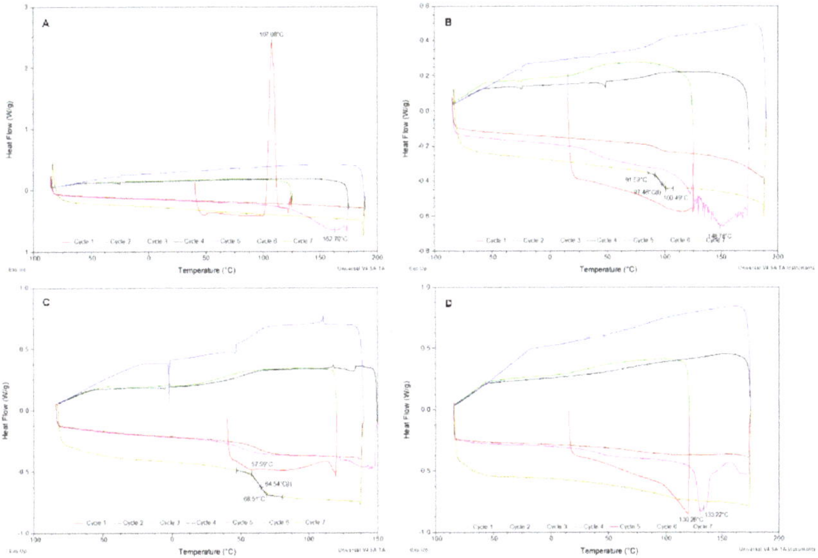

Figure 3. DSC thermograms of (**A**) [TMGH][ALN], (**B**) [TMGH]$_2$[ALN], (**C**) [C$_2$OHMIM][ALN] and (**D**) [DBNH][ALN].

The DSC thermogram of the RTIL [TMGH]$_2$[ALN] (Figure 3B) showed an endothermic signal at ca. 150 °C of the third cycle that could be assigned to a melting process. However, it is preceded by a glass transition at ca. 40 °C in the same cycle, meaning that the compound is in an amorphous state. So, the referred endothermic signal is most likely due to evaporation, consistent with the irregular shape of the curve caused by the formation of bubbles in the thick pasty compound. A similar observation can also be observed in [C_2OHMIM][ALN] (Figure 3C). From the set of eight compounds, [DBNH][ALN] is the only one that possesses a polymorphic structure, given by the two melting temperatures at 130.3 and 133.2 °C recorded in the DSC thermogram (Figure 3D). The remaining RTILs, more precisely [DBNH]$_2$[ALN] and [Ch]$_2$[ALN] show well-defined glass transitions confirming their amorphous nature at room temperature (Figures S29 and S33, respectively).

4. Solubility Studies

As expected, all OSILs were more soluble in water and saline solution at 37 °C than alendronic acid as well as its sodium salt. Figure 4 contains the data obtained from the solubility studies.

With the exception of the [DBNH]- and [C_2OHMIM]-containing ILs, the dianionic ALN–OSILs are equally or more soluble than the monoanionic siblings in the tested media. In further detail, it is noteworthy that [TMGH]$_2$[ALN], [DBNH][ALN] and [Ch]$_2$[ALN] are completely soluble in both water and saline solution in the tested conditions. Moreover, the [C_2OHMIM]-, [C_2OHMIM]$_2$- and [Ch]-based ALN–OSILs are also fully soluble in saline solution while the solubility in water was found to be between 392 and 918 times higher in comparison with the parent neutral drug, and between 77 and 198 times when compared with Na[ALN]. In addition, [DBNH]$_2$[ALN] presents full water solubility and a 1682-fold increase in solubility in saline solution. Finally, [TMGH][ALN] shows the lowest solubility in both media from the set of eight synthesized compounds.

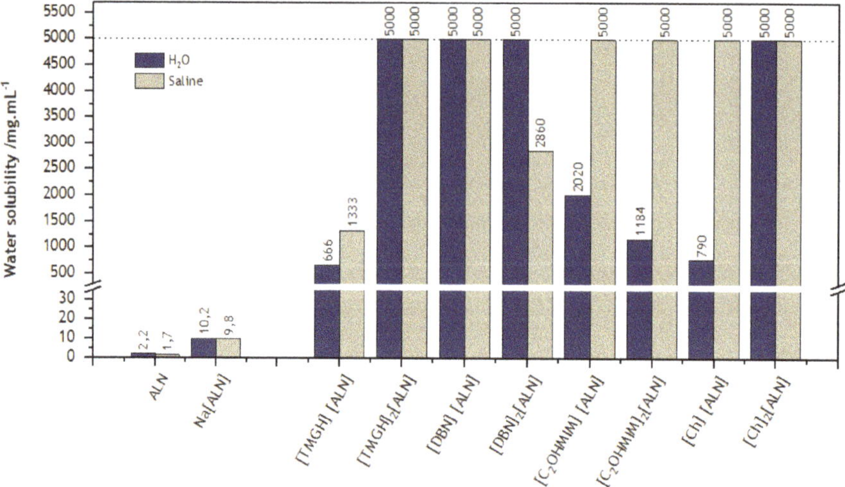

Figure 4. Solubility in water and saline solution at 37 °C of ALN, Na[ALN] and corresponding ALN–OSILs (detection limit is 5 g/mL, represented by the upper threshold).

5. Cytotoxicity on Human Cells

The cytotoxicity of the prepared ALN–OSILs was determined on human cells, by means of IC_{50} calculation. The analysis was performed on human gingival fibroblasts (GF) and on different cancer cell lines, namely T47D (breast), A549 (lung) and MG63 (osteosarcoma). The results obtained with the starting materials of the synthesis, as well as the prepared ALN–OSILs upon incubation for 24 h are presented in Table 2. The cytotoxic effect of the ALN-OSILs on the cancer cell lines was also evaluated after 72h of exposure, and the data is presented in Table 3. A comparison with the standard anticancer drug paclitaxel is presented for both exposure times.

Table 2. Cytotoxicity of the ALN–OSILs on GF cells and T47D, A549 and MG63 cell lines after 24 h.

	IC_{50}/mM			
IL	GF	T47D	A549	MG63
Paclitaxel	$1.91 \times 10^{-5} \pm 0.34 \times 10^{-5}$	$6.46 \times 10^{-6} \pm 0.57 \times 10^{-6}$	$4.08 \times 10^{-6} \pm 0.60 \times 10^{-6}$	$8.19 \times 10^{-6} \pm 1.03 \times 10^{-6}$
ALN	$3.17 \times 10^{-2} \pm 0.12 \times 10^{-2}$	$4.09 \times 10^{-3} \pm 0.12 \times 10^{-3}$ (*)	$8.10 \times 10^{-3} \pm 0.88 \times 10^{-3}$ (*)	$5.55 \times 10^{-4} \pm 0.79 \times 10^{-4}$ (*)
[TMGH]Cl	$1.47 \times 10^{-3} \pm 0.40 \times 10^{-3}$	$1.94 \times 10^{-4} \pm 0.35 \times 10^{-4}$	$4.98 \times 10^{-4} \pm 0.55 \times 10^{-4}$	$2.46 \times 10^{-4} \pm 0.81 \times 10^{-4}$
[DBNH]Cl	$3.80 \times 10^{-5} \pm 0.53 \times 10^{-5}$	$2.78 \times 10^{-4} \pm 0.19 \times 10^{-4}$	$9.13 \times 10^{-3} \pm 1.03 \times 10^{-3}$	$9.79 \times 10^{-8} \pm 1.76 \times 10^{-8}$
[Ch]Cl	(a)	(a)	$5.99 \times 10^{-3} \pm 0.72 \times 10^{-3}$	$3.47 \times 10^{-4} \pm 0.52 \times 10^{-4}$
[TMGH][ALN]	$7.19 \times 10^{-3} \pm 0.97 \times 10^{-3}$	$8.23 \times 10^{-3} \pm 0.99 \times 10^{-3}$	$2.54 \times 10^{-3} \pm 0.32 \times 10^{-3}$	$9.23 \times 10^{-3} \pm 1.12 \times 10^{-3}$
$[TMGH]_2$[ALN]	$1.11 \times 10^{-3} \pm 0.22 \times 10^{-3}$	$3.16 \times 10^{-5} \pm 0.25 \times 10^{-5}$	$5.14 \times 10^{-4} \pm 0.71 \times 10^{-4}$	$7.11 \times 10^{-5} \pm 0.85 \times 10^{-5}$
[DBNH][ALN]	$7.20 \times 10^{-2} \pm 0.55 \times 10^{-2}$	(a)	6.14 ± 0.87	(a)
$[DBNH]_2$[ALN]	$6.01 \times 10^{-5} \pm 0.94 \times 10^{-5}$	(a)	$3.12 \times 10^{-2} \pm 0.45 \times 10^{-2}$	1.02 ± 0.18
$[C_2OHMIM]$[ALN]	2.19 ± 0.32	$5.92 \times 10^{-3} \pm 0.69 \times 10^{-3}$ (*)	$2.10 \times 10^{-6} \pm 0.37 \times 10^{-6}$ (*)	$5.16 \times 10^{-5} \pm 0.67 \times 10^{-5}$ (*)
$[C_2OHMIM]_2$[ALN]	$4.07 \times 10^{-4} \pm 0.47 \times 10^{-4}$	$3.28 \times 10^{-6} \pm 0.40 \times 10^{-6}$ (*)	(a)	$7.84 \times 10^{-5} \pm 0.98 \times 10^{-5}$ (*)
[Ch][ALN]	$1.87 \times 10^{-1} \pm 0.32 \times 10^{-1}$	3.14 ± 0.43	$6.66 \times 10^{-2} \pm 0.89 \times 10^{-2}$ (*)	$4.10 \times 10^{-1} \pm 0.62 \times 10^{-1}$
$[Ch]_2$[ALN]	$4.92 \times 10^{-3} \pm 0.58 \times 10^{-3}$	$2.64 \times 10^{-3} \pm 0.38 \times 10^{-3}$	$5.41 \times 10^{-3} \pm 0.61 \times 10^{-3}$	$1.01 \times 10^{-2} \pm 0.14 \times 10^{-2}$

(a) Not determined in the tested concentration range. (*) Significantly lower than GF; $p < 0.05$.

Table 3. Cytotoxicity of the ALN–OSILs on GF cells and T47D, A549 and MG63 cell lines after 72 h.

	IC_{50}/mM			
IL	GF	T47D	A549	MG63
Paclitaxel	$5.66 \times 10^{-5} \pm 0.94 \times 10^{-5}$	$9.28 \times 10^{-6} \pm 1.02 \times 10^{-6}$	$7.83 \times 10^{-6} \pm 0.85 \times 10^{-6}$	$1.10 \times 10^{-6} \pm 1.25 \times 10^{-6}$
ALN	$9.26 \times 10^{-2} \pm 0.99 \times 10^{-2}$	$8.93 \times 10^{-3} \pm 1.53 \times 10^{-3}$ (*)	$1.46 \times 10^{-2} \pm 0.43 \times 10^{-2}$ (*)	$1.07 \times 10^{-3} \pm 0.26 \times 10^{-3}$ (*)
[TMGH][ALN]	$1.33 \times 10^{-2} \pm 0.51 \times 10^{-2}$	$1.84 \times 10^{-2} \pm 0.24 \times 10^{-2}$	$6.16 \times 10^{-3} \pm 0.94 \times 10^{-3}$	$1.64 \times 10^{-2} \pm 0.82 \times 10^{-2}$
$[TMGH]_2$[ALN]	$1.68 \times 10^{-3} \pm 0.27 \times 10^{-3}$	$1.87 \times 10^{-5} \pm 0.43 \times 10^{-5}$ (*)	$9.84 \times 10^{-4} \pm 0.67 \times 10^{-4}$	$1.67 \times 10^{-4} \pm 0.25 \times 10^{-4}$
[DBNH][ALN]	$2.55 \times 10^{-1} \pm 0.34 \times 10^{-1}$	(a)	0.99 ± 0.07	(a)
$[DBNH]_2$[ALN]	$1.65 \times 10^{-4} \pm 0.25 \times 10^{-5}$	(a)	$9.33 \times 10^{-2} \pm 1.00 \times 10^{-2}$	1.14 ± 0.21
$[C_2OHMIM]$[ALN]	4.54 ± 0.66	$1.14 \times 10^{-2} \pm 0.25 \times 10^{-2}$ (*)	$5.65 \times 10^{-6} \pm 0.46 \times 10^{-6}$ (*)	$9.57 \times 10^{-5} \pm 0.91 \times 10^{-5}$ (*)
$[C_2OHMIM]_2$[ALN]	$6.21 \times 10^{-4} \pm 0.67 \times 10^{-4}$	$8.83 \times 10^{-6} \pm 0.94 \times 10^{-6}$ (*)	(a)	$1.65 \times 10^{-4} \pm 0.26 \times 10^{-4}$ (*)
[Ch][ALN]	$7.33 \times 10^{-1} \pm 0.63 \times 10^{-1}$	4.86 ± 0.55	$7.84 \times 10^{-2} \pm 0.83 \times 10^{-2}$ (*)	$6.85 \times 10^{-1} \pm 0.85 \times 10^{-1}$
$[Ch]_2$[ALN]	$8.57 \times 10^{-3} \pm 0.87 \times 10^{-3}$	$4.87 \times 10^{-3} \pm 0.65 \times 10^{-3}$	$8.94 \times 10^{-3} \pm 1.34 \times 10^{-3}$	$8.33 \times 10^{-2} \pm 0.93 \times 10^{-2}$

(a) Not determined in the tested concentration range. (*) Significantly lower than GF, $p < 0.05$.

Globally, it was observed that the IC_{50} values determined at 72 h of incubation were higher than those obtained with a culture period of 24 h. The IC_{50} values obtained for paclitaxel in the tumor cell lines were the lowest ones, which is in line with its well-known cytotoxic potential. However, it showed no selectivity between healthy and cancer cells.

In the case of the ALN–OSILs, the obtained data clearly shows that the monoanionic OSILs elicited lower cytotoxicity than the corresponding dianionic versions for all cell types. Regarding gingival fibroblast cell cultures, the OSILs containing one unit of [DBNH], [C_2OHMIM] and [Ch] appeared to be less deleterious than alendronic acid. Comparatively, OSILs containing $[TMGH]_2$, [C_2OHMIM] and $[C_2OHMIM]_2$ seemed to be more cytotoxic to breast cancer T47D cells. In the case of lung cancer A549 cell line, high cytotoxic activity was observed in cell cultures supplemented with the OSILs containing $[TMGH]_2$, [C_2OHMIM] and [Ch]. Finally, osteosarcoma MG63 cell line seemed to be particularly sensitive to the OSILs containing $[TMGH]_2$, [C_2OHMIM] and $[C_2OHMIM]_2$.

Breast and lung cancers are often associated with bone osteolytic metastases [27–29] and also osteosarcoma, which are usually caused by disturbances in bone metabolism and increases in bone turnover rates [6,30,31]. Thus, the development of OSILs with high cytotoxicity towards the tested cancer cell types and containing an anti-resorbing molecule (alendronate) may represent a promising strategy for the development of new pharmacological tools to be used in those pathological conditions.

Overall, [C_2OHMIM][ALN] was found to be particularly active against lung cancer and osteosarcoma cell lines while retaining very low toxicity towards healthy cells. These enhanced biological properties, in addition to the absence of polymorphism in this monoanionic compound, suggest that this room temperature ionic liquid could be a very promising alendronic acid formulation.

6. Experimental Section

6.1. General Procedure (A) for the Synthesis of ALN–OSILs with Organic Superbases as Cations

To a dispersion of alendronic acid (400 mg, 1.61 mmol) in MeOH/H_2O (15 mL, 1:1) a methanolic solution of 1 or 2 molar equivalents of organic superbase (15 mg/mL) was added dropwise at room temperature under magnetic stirring. After reacting for 1 h, the solvent was evaporated and the desired product was dried under vacuo for 24 h.

6.2. General Procedure (B) for the Preparation of ALN–OSILs with Ammonium and Methylmidazolium Cations

The halide salts of the selected ammonium and methylimidazolium cations were dissolved in methanol and passed slowly through an anion-exchange column A-26(OH) (3 equivalents). The freshly formed methanolic solutions of the corresponding hydroxide salts (1 or 2 equivalents) were consequently added dropwise to alendronic acid (400 mg, 1.61 mmol) dispersed in H_2O under magnetic stirring at room temperature. After 1 h, the solvent of the clear solution was evaporated, and the desired product was dried under vacuo for 24 h.

Toxicity studies are described as supporting information.

Supplementary Materials: The following are available online at http://www.mdpi.com/1999-4923/12/3/293/s1, Materials, experimental procedures, compound characterization data. Figures S1–S16: NMR spectra of ALN-OSILs; Figures S17–S25: FTIR spectra of ALN and ALN-OSILs; Figures S26–S33: DSC thermograms of ALN-OSILs.

Author Contributions: Conceptualization, L.C.B. and J.C.-R.; methodology and investigation, S.T. and M.M.S.; validation, M.M.S., J.C.-R. and L.C.B; writing—Original draft preparation, M.M.S. and S.T.; writing—review and editing, M.M.S., J.C.-R. and L.C.B.; supervision, M.H.F., J.C.-R. and L.C.B.; funding acquisition, M.H.F., J.C.-R. and L.C.B. All authors have read and agree to the published version of the manuscript.

Funding: This work was supported by the Associate Laboratory for Green Chemistry LAQV which is financed by national funds from FCT/MCTES (UIDB/50006/2020). The authors also thank Fundação para a Ciência e Tecnologia for the projects PTDC/QUI-QOR/32406/2017, PEst-C/LA0006/2013 and RECI/BBBBQB/0230/2012), as well as one contract under Investigador FCT (L. C. Branco). The authors also acknowledge the Solchemar company.

Acknowledgments: The authors also thank Madalena Dionísio for their support with the DSC analysis.

Conflicts of Interest: The authors declare no conflict of interest.

Abbreviations

IL	ionic liquid
ALN	alendronic acid
ALN–OSILs	alendronic acid-based ionic liquids and organic salts
RTILs	Room Temperature Ionic Liquids
API–OSILs	Active Pharmaceutical Ingredient Ionic Liquids and Organic Salts
TMG	1,1,3,3-tetramethylguanidine
DBN	1,5-diazabicyclo(4.3.0)non-5-ene
Ch	choline
C_2OHMIM	1-(2-hydroxyethyl)-3-methylimidazolium

References

1. Shinkai, I.; Ohta, Y. New drugs—Reports of new drugs recently approved by the FDA: Alendronate. *Bioorg. Med. Chem.* **1996**, *4*, 3–4. [CrossRef]
2. Han, H.K.; Shin, H.J.; Ha, D.H. Improved oral bioavailability of alendronate via the mucoadhesive liposomal delivery system. *Eur. J. Pharm. Sci.* **2012**, *46*, 500–507. [CrossRef] [PubMed]
3. Iwamoto, J.; Takeda, T.; Sato, Y. Efficacy and safety of alendronate and risedronate for postmenopausal osteoporosis. *Curr. Med. Res. Opin.* **2006**, *22*, 919–928. [CrossRef] [PubMed]
4. Brufsky, A.M. Bisphosphonates, bone, and breast cancer recurrence. *Lancet* **2015**, *386*, 2. [CrossRef]
5. Mathew, A.; Brufsky, A.M. The Use of Adjuvant Bisphosphonates in the Treatment of Early-Stage Breast Cancer. *Clin. Adv. Hematol. Oncol.* **2014**, *12*, 8.
6. Avnet, S.; Longhi, A.; Salerno, M.; Halleen, J.M.; Perut, F.; Granchi, D.; Ferrari, S.; Bertoni, F.; Giunti, A.; Baldini, N. Increased osteoclast activity is associated with aggressiveness of osteosarcoma. *Int. J. Oncol.* **2008**, *33*, 1231–1238. [CrossRef]
7. Hough, W.L.; Smiglak, M.; Rodriguez, H.; Swatloski, R.P.; Spear, S.K.; Daly, D.T.; Pernak, J.; Grisel, J.E.; Carliss, R.D.; Soutullo, M.D.; et al. The third evolution of ionic liquids: Active pharmaceutical ingredients. *New J. Chem.* **2007**, *31*, 1429–1436. [CrossRef]
8. Egorova, K.S.; Gordeev, E.G.; Ananikov, V.P. Biological Activity of Ionic Liquids and Their Application in Pharmaceutics and Medicine. *Chem. Rev.* **2017**, *117*, 7132–7189. [CrossRef]
9. Ferraz, R.; Branco, L.C.; Prudencio, C.; Noronha, J.P.; Petrovski, Z. Ionic Liquids as Active Pharmaceutical Ingredients. *ChemMedChem* **2011**, *6*, 975–985. [CrossRef]
10. Marrucho, I.M.; Branco, L.C.; Rebelo, L.P.N. Ionic Liquids in Pharmaceutical Applications. *Annu. Rev. Chem. Biomol. Eng.* **2014**, *5*, 527–546. [CrossRef]
11. Smiglak, M.; Pringle, J.M.; Lu, X.; Han, L.; Zhang, S.; Gao, H.; MacFarlane, D.R.; Rogers, R.D. Ionic liquids for energy, materials, and medicine. *Chem. Comm.* **2014**, *50*, 9228–9250. [CrossRef] [PubMed]
12. Shamshina, J.L.; Kelley, S.P.; Gurau, G.; Rogers, R.D. Chemistry: Develop ionic liquid drugs. *Nature* **2015**, *528*, 188–189. [CrossRef] [PubMed]
13. Cojocaru, O.A.; Bica, K.; Gurau, G.; Narita, A.; McCrary, P.D.; Shamshina, J.L.; Barber, P.S.; Rogers, R.D. Prodrug ionic liquids: Functionalizing neutral active pharmaceutical ingredients to take advantage of the ionic liquid form. *MedChemComm* **2013**, *4*, 559–563. [CrossRef]
14. Cherukuvada, S.; Nangia, A. Polymorphism in an API ionic liquid: Ethambutol dibenzoate trimorphs. *CrystEngComm* **2012**, *14*, 7840–7843. [CrossRef]
15. Santos, M.M.; Raposo, L.R.; Carrera, G.V.S.M.; Costa, A.; Dionísio, M.; Baptista, P.V.; Fernandes, A.R.; Branco, L.C. Ionic Liquids and Salts from Ibuprofen as Promising Innovative Formulations of an Old Drug. *ChemMedChem* **2019**, *14*, 907–910. [CrossRef]
16. Carrera, G.V.S.M.; Santos, M.M.; Costa, A.; Rebelo, L.P.N.; Marrucho, I.M.; Ponte, M.N.; Branco, L.C. Highly water soluble room temperature superionic liquids of APIs. *New J. Chem.* **2017**, *41*, 6986–6990.
17. Ferraz, R.; Branco, L.C.; Marrucho, I.; Araújo, J.; da Ponte, M.N.; Prudêncio, C.; Noronha, J.P.; Petrovski, Z. Development of Novel Ionic Liquids-APIs based on Ampicillin derivatives. *Med. Chem. Comm.* **2012**, *3*, 494–497. [CrossRef]

18. Florindo, C.; Araujo, J.M.M.; Alves, F.; Matos, C.; Ferraz, R.; Prudencio, C.; Noronha, J.P.; Petrovski, Z.; Branco, L.; Rebelo, L.P.N.; et al. Evaluation of solubility and partition properties of ampicillin-based ionic liquids. *Int. J. Pharm.* **2013**, *456*, 553–559. [CrossRef]
19. Ferraz, R.; Teixeira, V.; Rodrigues, D.; Fernandes, R.; Prudencio, C.; Noronha, J.P.; Petrovski, Z.; Branco, L.C. Antibacterial activity of Ionic Liquids based on ampicillin against resistant bacteria. *RSC Adv.* **2014**, *4*, 4301–4307. [CrossRef]
20. Ferraz, R.; Silva, D.; Dias, A.R.; Dias, V.; Santos, M.M.; Pinheiro, L.; Prudêncio, C.; Noronha, J.P.; Petrovski, Z.; Branco, L.C. Synthesis and Antibacterial Activity of Ionic Liquids and Organic Salts based on Penicillin G and Amoxicillin hydrolysate derivatives against Resistant Bacteria. *Pharmaceutics* **2020**, *12*, 221. [CrossRef]
21. Florindo, C.; Costa, A.; Matos, C.; Nunes, S.L.; Matias, A.N.; Duarte, C.M.M.; Rebelo, L.P.N.; Branco, L.C.; Marrucho, I.M. Novel organic salts based on fluoroquinolone drugs: Synthesis, bioavailability and toxicological profiles. *Int. J. Pharm.* **2014**, *469*, 179–189. [CrossRef] [PubMed]
22. Teixeira, S.; Santos, M.M.; Ferraz, R.; Prudêncio, C.; Fernandes, M.H.; Costa-Rodrigues, J.; Branco, L.C. A Novel Approach for Bisphosphonates: Ionic Liquids and Organic Salts from Zoledronic Acid. *ChemMedChem* **2019**, *14*, 1767–1770. [CrossRef] [PubMed]
23. Frade, R.; Rosatella, A.A.; Marques, C.S.; Branco, L.C.; Kulkarni, P.S.; Mateus, N.M.M.; Afonso, C.A.M.; Duarte, C.M.M. Toxicological evaluation on human colon carcinoma cell line (CaCo-2) of ionic liquids based on imidazolium, guanidinium, ammonium, phosphonium, pyridinium and pyrrolidinium cations. *Green Chem.* **2009**, *11*, 1660–1665. [CrossRef]
24. NIST Chemistry Web Book. Available online: http://webbook.nist.gov (accessed on 5 January 2020).
25. Mezzetta, A.; Łuczak, J.; Woch, J.; Chiappe, C.; Nowicki, J.; Guazzelli, L. Surface active fatty acid ILs: Influence of the hydrophobic tail and/or the imidazolium hydroxyl functionalization on aggregates formation. *J. Mol. Liq.* **2019**, *289*, 111155. [CrossRef]
26. Afergan, E.; Najajreh, Y.; Gutman, D.; Epstein, H.; Elmalak, O.; Golomb, G. 31P-NMR and Differential Scanning Calorimetry Studies for Determining Vesicle's Drug Physical State and Fraction in Alendronate Liposomes. *J. Bioanal. Biomed.* **2010**, *2*, 125–131. [CrossRef]
27. Kan, C.; Vargas, G.; Pape, F.L.; Clézardin, P. Cancer Cell Colonisation in the Bone Microenvironment. *Int. J. Mol. Sci.* **2016**, *17*, 1674. [CrossRef]
28. Shemanko, C.S.; Cong, Y.; Forsyth, A. What Is Breast in the Bone? *Int. J. Mol. Sci.* **2016**, *17*, 1764. [CrossRef]
29. Costa-Rodrigues, J.; Moniz, K.A.; Teixeira, M.R.; Fernandes, M.H. Variability of the paracrine-induced osteoclastogenesis by human breast cancer cell lines. *J. Cell. Biochem.* **2012**, *113*, 1069–1079. [CrossRef]
30. Costa-Rodrigues, J.; Fernandes, A.; Fernandes, M.H. Reciprocal osteoblastic and osteoclastic modulation in co-cultured MG63 osteosar-coma cells and human osteoclast precursors. *J. Cell. Biochem.* **2011**, *112*, 3704–3713. [CrossRef]
31. Costa-Rodrigues, J.; Teixeira, C.A.; Fernandes, M.H. Paracrine-mediated osteoclastogenesis by the osteosarcoma MG63 cell line: Is RANKL/RANK signaling really important? *Clin. Exp. Metastas.* **2011**, *28*, 505–514. [CrossRef]

Article

Synthesis and Antibacterial Activity of Ionic Liquids and Organic Salts Based on Penicillin G and Amoxicillin Hydrolysate Derivatives against Resistant Bacteria

Ricardo Ferraz [1,2,*,†], Dário Silva [3,†], Ana Rita Dias [1,2], Vitorino Dias [1], Miguel M. Santos [3], Luís Pinheiro [3], Cristina Prudêncio [1,4], João Paulo Noronha [3], Željko Petrovski [3,*] and Luís C. Branco [3,*]

1 Ciências Químicas e das Biomoléculas (CQB) e Centro de Investigação em Saúde e Ambiente (CISA), Escola Superior de Saúde do Instituto Politécnico do Porto, 4400-330 Porto, Portugal; anaritadias3@gmail.com (A.R.D.); solchemar@gmail.com (V.D.); cps@estsp.ipp.pt (C.P.)

2 LAQV-REQUIMTE, Departamento de Química e Bioquímica, Faculdade de Ciências, Universidade do Porto, Rua do Campo Alegre 687, 4169-007 Porto, Portugal

3 LAQV-REQUIMTE, Departamento de Química, Faculdade de Ciências e Tecnologia da Universidade Nova de Lisboa, 2829-516 Caparica, Portugal; dmv.silva@campus.fct.unl.pt (D.S.); miguelmsantos@fct.unl.pt (M.M.S.); l.pinheiro@campus.fct.unl.pt (L.P.); jpnoronha@fct.unl.pt (J.P.N.)

4 i3S, Instituto de Inovação e Investigação em Saúde, Universidade do Porto, 4099-002 Porto, Portugal

* Correspondence: ricardoferraz@eu.ipp.pt (R.F.); z.petrovski@fct.unl.pt (Ž.P.); l.branco@fct.unl.pt (L.C.B.)

† These two authors contribute equally to this work.

Received: 31 December 2019; Accepted: 19 February 2020; Published: 2 March 2020

Abstract: The preparation and characterization of ionic liquids and organic salts (OSILs) that contain anionic penicillin G [secoPen] and amoxicillin [*seco*-Amx] hydrolysate derivatives and their in vitro antibacterial activity against sensitive and resistant *Escherichia coli* and *Staphylococcus aureus* strains is reported. Eleven hydrolyzed β-lactam-OSILs were obtained after precipitation in moderate-to-high yields via the neutralization of the basic ammonia buffer of antibiotics with different cation hydroxide salts. The obtained minimum inhibitory concentration (MIC) data of the prepared compounds showed a relative decrease of the inhibitory concentrations (RDIC) in the order of 100 in the case of [C_2OHMIM][*seco*-Pen] against sensitive *S. aureus* ATCC25923 and, most strikingly, higher than 1000 with [C_{16}Pyr][*seco*-Amx] against methicillin-resistant *Staphylococcus aureus* (MRSA) ATCC 43300. These outstanding in vitro results showcase that a straightforward transformation of standard antibiotics into hydrolyzed organic salts can dramatically change the pharmaceutical activity of a drug, including giving rise to potent formulations of antibiotics against deadly bacteria strains.

Keywords: active pharmaceutical ingredients-ionic liquids and organic salts (API-OSILs); penicillin G and amoxicillin hydrolysate derivatives; sensitive bacteria; resistant bacteria

1. Introduction

Bacterial resistance to antibiotics has been increasing in Europe over the last few years [1–3]. New classes of antibiotics have not been introduced recently [4–7], and, thus, more resistances to old drugs are developing daily [8–10]. Recent efforts and huge investments being made in this field by big pharma companies such as GlaxoSmithKline, Merck, Pfizer and Wyeth [3–5,11,12] have had disappointing returns from their R&D departments, including clinical trials. This is a significant factor to allocate anti-infective R&D resources into other fields of investigation and thus remain highly competitive [3–5,13]. Considering the disappointing results on genomics and the exodus of big pharma, the problem of bacteria resistance has continued to evolve, reaching alarming dimensions [3,8,10].

For the last 12 years, organic salts and ionic liquids (OSILs) from active pharmaceutical ingredients (APIs), or simply API-OSILs [14–22], have been studied at the academic level [14,18,19,23–25]. Ionic liquids (ILs) are salts with melting points below 100 °C (some of them are liquid at room temperature) that result from the pairing of organic cations with organic and inorganic anions [14,20,23]. When the melting point is above 100 °C, these compounds are simply designated by organic salts [24]. Nowadays, there is a significant increase in the scope of both the physical and chemical properties of OSILs [19,26–29], and, thus, their application in several topics of science and technology is currently being studied [15–18,20,23,26].

In the case of API-OSILs, it is known that the interaction between an ionic API with selected counter-ions may significantly improve the pharmaceutical activity of the former [17,30–32]. In addition, this combination may also boost the stability and solubility of the API in physiological media, as well as enhance the bioavailability and modify the pharmacokinetics and delivery mode of the drug [21,27–29,33–35]. Consequently, the biopharmaceutics drug classification system (BCS) for API-ILs can be significantly modified in comparison with the parent drugs [24], meaning that this new salt of the old API can be treated as a new chemical entity and thus be independently patented [17,26,30,36]. Furthermore, the polymorphism of a given API can be severely mitigated or even eliminated if it becomes liquid, hence tackling one of the most important problems in the pharmaceutical industry that can dramatically alter a drug's solubility and dosages [26,30,37–40]. In fact, solid forms of drugs can suffer from several limitations such as low solubility, polymorphic conversion, and low bioavailability [20,26,36,41].

The inherent properties of ILs could be of extreme importance to overcome such difficulties of solid form drugs [20,36,41]. Recent works have shown that API-OSILs possess many attractive properties when compared to conventional drugs [14,16,20,23]. Our group recently studied the relevant pharmacological properties of ampicillin- and primaquine-based API-OSILs such as water solubility, the octanol–water partition coefficient, the hexadecylphosphocholine (HDPC) micelle–water partition coefficient, and critical micelle concentration [14–17,30]. In the case of ampicillin-based API-OSILs, the data are clearly consistent with a greater potential of API-ILs in comparison with the parent API, specifically regarding their solubility in water, as well as more specific properties such as membrane affinity and permeation. In fact, the accurate selection of the organic cation allows for the fine-tuning of some important physical and thermal properties like water solubility, membrane permeation, melting point, and thermal stability [17]. In another study, we found that primaquine API-OSILs had a particular affinity to intercalate negatively-charged lipid bilayers (membrane models of *Plasmodium* infected erythrocyte) and also, to a lesser extent, zwitterionic lipid bilayers (membrane models of healthy cells), in comparison with the parent drug [14,42].

A large quantity of recent communications and reviews have referred to the toxicity and activity of ILs against microorganisms and cell cultures, especially antimicrobial activity, and as novel forms of bioactive materials and as drug delivery systems [14,20,43–46]. Recently, OSILs have been studied to fight multi-drug resistance [16,47] and as agents for microbial biofilms [32,45,48–51], and they have shown a potent, broad spectrum activity against a variety of clinically significant microbial pathogens, including methicillin-resistant *Staphylococcus aureus* (MRSA) [32,49,52]. The 2011 outbreak of multi drug-resistant *Escherichia coli* O104 in Germany as well as as Gram-negative Enterobacteriaceae due to presence of the New Delhi metallo-β-lactamase [53,54] are examples of an increasingly documented major public health problem.

Therefore, there is an increasing demand to develop new drugs to address multi-resistant infections and to develop more efficient tools so that new resistances are not developed. In this context, the results up to this point given by API-OSILs have followed these demands [20].

Thus, following our work on ampicillin [16–18] and fluoroquinolone-based API-OSILs [30], we herein describe the synthesis of ILs based on amoxicillin and penicillin G through a synthetic strategy that was optimized by us in the past for the preparation of ampicillin-OSILs [18]. The synthetic method consists on the deprotonation of an API with different hydroxides of organic cations in an ammonium buffer (buffer-controlled reaction procedure) [18]. In all cases, the antibiotic API is combined as an anion with organic cations that contain imidazolium, ammonium, phosphonium and pyridinium structures.

2. Results and Discussion

2.1. Chemistry

Ionic liquids and organic salts based on the ammonia hydrolysate anion of penicillin G and amoxicillin (α-amide of benzyl penicilloic acid and of amoxicillin penicilloic acid, respectively), abbreviated here as [*seco*-Pen] and [*seco*-Amx], respectively, were prepared by an ammonia buffer reaction procedure that was recently developed by us for the synthesis of ampicillin API-OSILs [18]. Our original idea was to test a new method for preparation of API-OSILs based on parent penicillin and amoxicillin in the anionic form. However, due to ammonolysis (β-lactam ring opening with the formation of an amide group) of amoxicillin and penicillin G under the employed reaction conditions, ionic liquids and organic salts of [*seco*-Pen] and [*seco*-Amx] anions were prepared instead (see Figure 1). The prefix *seco* (Latin verb secare) is used in antibiotics nomenclature [55], and it means to cut. Two penicilloic acids of penicillin G and amoxicillin are already well known in the literature [56–58], and some of their stable amides—products of the amino- and ammonolysis of β-lactam ring of parent antibiotics—have also already been described [59–61].

While amoxicillin (Amx) was used as provided (in trihydrate form), penicillin G, which was supplied as a potassium salt ([K][Pen]), was previously converted to the corresponding ammonium salt $[NH_4]$[Pen] by following the procedure of Brown et al. [62]. The experimental procedure consisted on the reaction of hydroxide of the selected cations with amoxicillin and penicillin ammonium salt. Halide (chloride, bromide) salts of selected organic cations were converted into the corresponding hydroxides on Amberlite resin (OH form), and the highly basic solution that was obtained was then added to the solution of the API in an aqueous ammonia buffer in order to provide the compounds [18]. [Na][*seco*-Amx] was prepared through a reaction with sodium hydroxide (instead of organic hydroxide), while a [K][*seco*-Pen] derivative was directly obtained by ammonolysis from [K][Pen].

Hydrolyzed amoxicillin (*seco*-Amx) and penicillin G (*seco*-Pen) derivatives in the anionic form were combined with the following organic cations (see Figure 1): 1-ethyl-3-methylimidazolium [EMIM], 1-hydroxy-ethyl-3-methylimidazolium [C_2OHMIM], (2-hydroxyethyl)trimethylammonium [Choline], tetraethylammonium [TEA], cetylpyridinium [C_{16}Pyr] and trihexyltetradecylphosphonium [$P_{6,6,6,14}$]. These cations were selected due to their low toxicity, except for [$P_{6,6,6,14}$], which was chosen because it usually produces room temperature ionic liquids (RTILs). Cetylpyridinium chloride is also already used in some types of mouthwashes and toothpastes [53], although it is irritant in higher concentrations [54]. For control purposes in the biological activity studies, we also prepared the corresponding sodium and potassium salts of the hydrolyzed antibiotics by following the same synthetic procedure.

The isolation of these compounds was performed similarly to the previously reported ampicillin-based API-OSILs [18]. Briefly, the excess reactant was filtered-off after crystallization from acetonitrile/methanol (9:1), and this was followed by solvent evaporation. Table 1 shows the reaction yields, physical states and melting points of the prepared compounds. From the eleven synthesized API-OSILs, two were organic salts and nine were ionic liquids, including four RTILs.

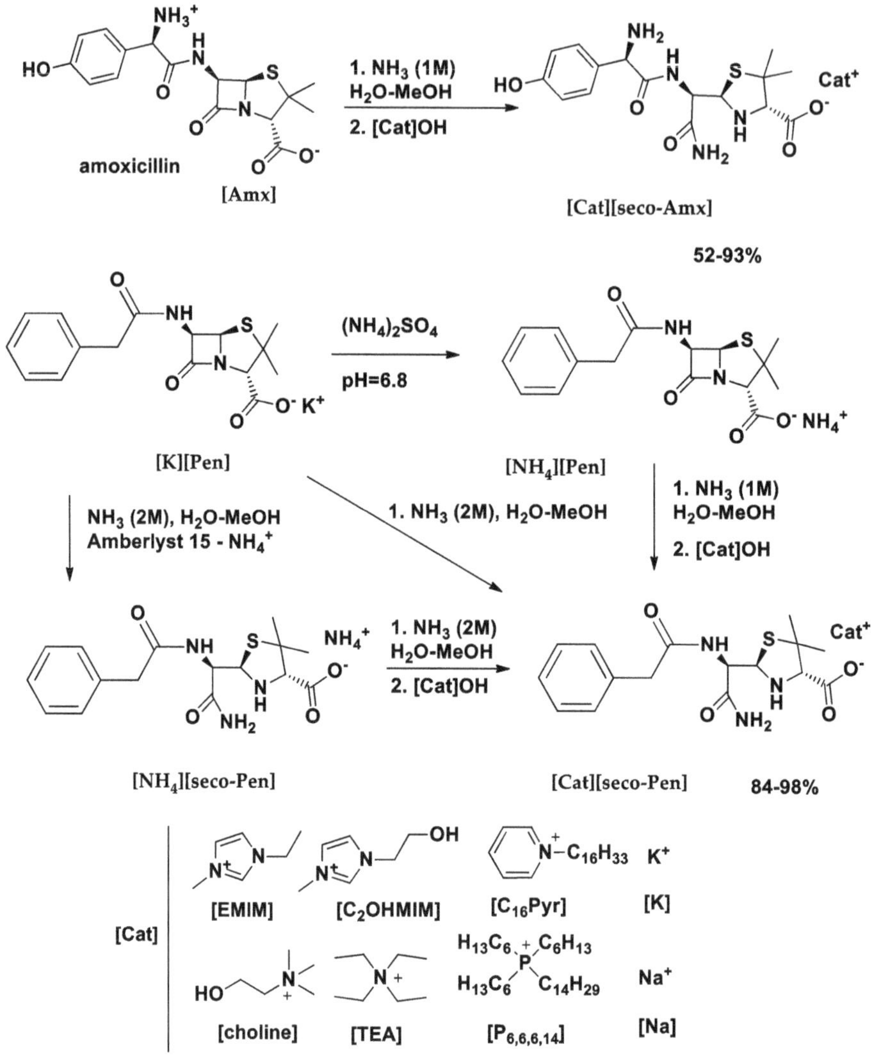

Figure 1. Schematic synthetic methodology for the preparation of metallic active pharmaceutical ingredients (API) salts and [*seco*-Amx] and [*seco*-Pen] ionic liquids and organic salts (OSILs).

The strict 1:1 cation–anion stoichiometry of the prepared API-OSILs, as well as the structural integrity of both components, was confirmed by ^{1}H NMR spectroscopy. Further characterization was performed by ^{13}C NMR and FTIR spectroscopies, as well as specific rotation, elemental analysis, and mass spectrometry. The ammonolysis of the β-lactam rings of the original APIs was confirmed by mass spectrometry. In the acquired electrospray ionization mass spectra (ESI-MS) spectra of all analyzed API-OSILs in the negative mode, the base peak corresponded to $[M + 17]^-$ (*m/z*), which was consistent with the β-lactam ring opening with the consequent formation of an amide group and a secondary amine at the thiazolidine group.

Table 1. Yield, physical state, and melting point of the synthesized API-OSILs.

Compound	Yield	Physical State	Melting Point/°C
[EMIM][*seco*-Amx]	77%	Yellow solid	84–86
[C_2OHMIM][*seco*-Amx]	60%	Yellow solid	109–111
[$P_{6,6,6,14}$][*seco*-Amx]	92%	Yellow viscous liquid	-
[C_{16}Pyr][*seco*-Amx]	94%	Yellow solid	96–98
[Choline][*seco*-Amx]	93%	Yellow solid	143–144
[Na][*seco*-Amx]	96%	Yellow solid	137–139
[EMIM][*seco*-Pen]	81%	Colorless viscous liquid	
[C_2OHMIM][*seco*-Pen]	83%	Yellow solid	48–50
[Choline][*seco*-Pen]	95%	Yellow solid	69–71
[$P_{6,6,6,14}$][*seco*-Pen]	97%	Yellow viscous liquid	-
[C_{16}Pyr][*seco*-Pen]	89%	Yellow solid	76–78
[TEA][*seco*-Pen]	90%	Yellow viscous liquid	-
[K][*seco*-Pen]	97%	White solid	193–195

* This is a outside table footnote.

2.2. Biological Activity

In the present study, all prepared compounds were tested against several sensitive and resistant Gram-positive and Gram-negative bacteria strains: *Staphylococcus aureus* ATCC 25923, *Escherichia coli* ATCC 25922, methicillin resistant *Staphylococcus aureus* (MRSA ATCC 43300) and *E. coli* CTX M2 and CTX M9.

The minimum inhibitory concentration (MIC) values were determined from three assays in triplicate by the broth micro dilution method in a 96-well microtiter plate by using tryptic soy broth (TSB) and adapted methodology from the Clinical Laboratory Standard Institute (CLSI) [62]. The strains were grown individually on tryptic soy agar for 24 h at 37 °C prior to each antibacterial test. Preceding MIC determination, each inoculum density was adjusted in TSB to 0.5 McFarland standards with a photometric device [62]. This resulted in a suspension that contained approximately 1×10^8 to 2×10^8 colony forming units (CFUmL^{-1}) for *E. coli* ATCC25922® [60]. A similar approach was used for the other strains. Then, 0.5 μL of the suspension was added to each well to have 5000–25,000 CFUmL^{-1}. Bacteria were exposed to API-OSIL concentrations of 5, 2.5, 0.5, 0.05, 0.005, 0.0005, and 0.00005 mM. All compounds were dissolved in water except for OSILs that contained [$P_{6,6,6,14}$] and [C_{16}Pyr] cations, which were diluted in 1% dimethyl sulfoxide (DMSO). Their activity was determined in aqueous media, and the results of their activity were compared with bacteria that had been grown in TSB broth in the presence of 1% DMSO. The MIC for each compound was recorded as lowest molar concentration, showing no turbidity after 24 h of incubation at 37 °C [63,64]. The presence of turbidity was an indication of microbial growth, and the corresponding concentration of antibacterial agent was considered ineffective. The pharmacological activity of the prepared compound was then compared to the parent commercial API in terms of the relative decrease of inhibitory concentration (RDIC) as described earlier [16]. Herein, the RDIC value was calculated by dividing the MIC of the commercial API (penicillin G potassium salt or amoxicillin trihydrate) by the MIC of the corresponding synthesized compound.

2.3. Studies on Sensitive Bacteria

Table 2 shows data from the bioactivity study of the prepared compounds on *S. aureus* and *E. coli* sensitive strains.

The data gathered in Table 2 show that, on a first approach, the hydrolyzed salts of both Amx and Pen lost all antimicrobial activity against the tested sensitive strains. In fact, this was true for the vast majority of prepared OSILs. It has been a consistently and firmly established belief that the antimicrobial activity of β-lactam antibiotics relies on the sacrificial action of these functional groups [65] on transpeptidases that are responsible for the last cross-linking step of peptidoglycan synthesis in the bacterial cell wall. For that purpose, the transpeptidase active site (Ser residues in the case of D, and D- and Cys residues in the case of L,D-transpeptidases) nucleophilically attacks the nucleophilicity the carbonyl of the β-lactam ring, resulting in its opening of the ring and the irreversible formation of a covalent and stable acyl-enzyme complex [66,67]. However, this premise is widely accepted but poorly demonstrated. Recently, it was shown that even the acylation reactions of Cys residue of L,D-transpeptidases can be reversible, thus leading to limited antibacterial activity [68].

Table 2. Minimum inhibitory concentrations (mM) and relative decrease of inhibitory concentrations (RDIC) of the new compounds that were produced on sensitive strains.

Compound	*S. aureus* ATCC25923	*RDIC*	*E. coli* ATCC25922	*RDIC*
[EMIM][*seco*-Amx]	5.0	0.01	2.5	0.002
[C_2OHMIM][*seco*-Amx]	0.050	1	5.0	0.001
[$P_{6,6,6,14}$][*seco*-Amx]	>5.0	<0.01	0.5	0.01
[C_{16}Pyr][*seco*-Amx]	>5.0	<0.01	0.050	0.1
[Choline][*seco*-Amx]	>5.0	<0.01	>5.0	<0.001
Na[*seco*-Amx]	>5.0	<0.01	>5.0	<0.001
Amx	0.050	1	0.005	1
[EMIM][*seco*-Pen]	>5.0	<0.1	>5.0	<0.1
[C_2OHMIM][*seco*-Pen]	**0.005**	**100**	>5.0	<0.1
[Choline][*seco*-Pen]	>5.0	<0.1	5.0	0.1
[$P_{6,6,6,14}$][*seco*-Pen]	>5.0	<0.1	>5.0	<0.1
[C_{16}Pyr][*seco*-Pen]	>5.0	<0.1	>5.0	0.1
[TEA][*seco*-Pen]	>5.0	<0.1	**0.050**	**10**
K[*seco*-Pen]	>5.0	<0.1	>5.0	<0.1
K[Pen]	0.500	1	0.500	1
[EMIM][Br]	0.05	—	>5	—
[C_2OHMIM][Cl]	>5.0	—	5.0	—
[$P_{6,6,6,14}$][Cl]	2.5	—	2.5	—
[C_{16}Pyr][Cl]	0.5	—	0.5	—
[Choline][Cl]	2.5	—	>5.0	—
[TEA][Br]	2.5	—	>5.0	—

The same result was obtained for the majority of the prepared salts, irrespective of the cation polarity, resembling an ion trapping effect [69], i.e., ionic compounds such as [*seco*-Amx] and [*seco*-Pen] and their conjugated acids are subject to a variety of processes, such as dissociation, electrical interactions with organic matter, and changes in their partitioning in hydrophobic/hydrophilic media. These processes depend on pH, ionic strength, their polarity, and pK_a, and they ultimately lead to their accumulation in certain zones of the bacterial cell, i.e., the ion-trap effect. On the other hand, highly polar cations are more prone to closely interact with them in anionic form, anchoring them in the polar solution. This effect has also been seen by us with ampicillin-based API-OSILs against sensitive Gram-positive and Gram-negative species [16] and seem to adversely affect the activity of those compounds. In the case of highly hydrophobic cations, the micelles of API-OSILs may be formed, thus reducing their antimicrobial activity [70].

The only observed exceptions were [C_2OHMIM][*seco*-Amx], [C_2OHMIM][*seco*-Pen], and [TEA][*seco*-Pen]. While the first one showed no advantage over Amx (RDIC = 1) against *S. aureus*, the second and third *seco*-Pen OSILs recorded RDICs of 100 (*S. aureus*) and 10 (*E. coli*), respectively.

While inactive in the Cl^- salt form, the $[C_2OHMIM]^+$ cation was the only selected cationic entity that allowed for the enhancement of the antimicrobial activity of both *seco*-Amx and *seco*-Pen against this sensitive *S. aureus* strain. These results most probably come from specific intermolecular interactions between the cation and the anion, thereby enabling the deactivating interactions of crucial PBPs. PBP:API-OSILs interaction studies will be conducted in the future and published accordingly.

In the case of the sensitive Gram-negative *E. coli* strain, its outer membrane can hinder drug delivery, as proposed above. In fact, the activity of [*seco*-Amx] and [*seco*-Pen] towards this strain did not seem to be enhanced by the combination with [C_2OHMIM] or with any of the other cations, with the exception of [TEA] (RDIC = 10). In truth, [TEA][*seco*-Pen] was found to be ten times more effective than the parent K[Pen]. These results are probably related with augmented hindrance of the porin entrance and/or uncompetitive phase transfer delivery through the outer membrane in comparison with the free antibiotic.

Similar results were also recorded in a past study regarding ampicillin-based API-ILs with sensitive Gram-positive and Gram-negative bacteria [16]. Recent results in the literature regarding API-OSILs as antibiotics against sensitive bacteria have referred to discrete interactions with the bacteria cell wall or membrane [71–79]. In particular, pyridyl cationic-modified benzylidene cyclopentanone photosensitizers (PSs) that were developed by Wu et al. [77] showed that Gram-positive bacteria are more sensitive than Gram-negative bacteria to photodynamic therapy because their walls are more porous and all types of PSs can readily diffuse through them. In contrast, Gram-negative bacteria possess an additional negatively charged outer layer that serves as a permeability barrier, so neutral and anionic PSs often fail to effectively inactivate Gram-negative bacteria, while cationic PSs can still strongly bind to their outer membrane and damage their integrity. In addition, antimicrobial studies that were supported by FTIR spectroscopy experiments revealed that nalidixic acid salts with particular ammonium cations exhibit enhanced antimicrobial activity against six different Gram-negative *Salmonella* species and two nalidixic acid-resistant *S. typhimurium* strains by displaying different modes of action towards proteins, carbohydrates, and lipids within the cell membrane [78]. One final example concerning FTIR bioassays revealed that hydrophobic *N*-alkyltropinium bromide surfactants preferably interact with the fatty acids and amide groups within the cell envelope of Gram-negative *E. coli* and with the peptidoglycan multilayer of the Gram-positive *Listeria innocua* cells [79]. The interaction of the ILs based on penicillin and amoxicillin, as well as [*seco*-Pen] and [*seco*-Amx] anions with the cell membrane of Gram-positive and Gram-negative bacteria strains, will be studied soon and published accordingly.

2.4. Studies on Resistant Bacteria

The prepared OSILs from hydrolyzed amoxicillin and penicillin antibiotics were also studied against resistant Gram-negative *E. coli* strains CTX M9 and CTX M2, as well as the methicillin-resistant *S. aureus* ATCC 43,300 (see Table 3).

Table 3. Minimum inhibitory concentrations (mM) and and relative decrease of inhibitory concentration (RDIC) of the new compounds that were produced on resistant strains.

Compound	*E. coli* CTX M9	*RDIC*	*E. coli* CTX M2	*RDIC*	MRSA ATCC 43300	*RDIC*
[EMIM][*seco*-Amx]	>5	-	>5	-	>5	-
[C_2OHMIM][*seco*-Amx]	>5	-	> 5	-	5	>1
[$P_{6,6,6,14}$][*seco*-Amx]	0.05	>100	1.0	>5	> 5	-
[C_{16}Pyr][*seco*-Amx]	0.05	>100	0.05	>100	0.005	>1000
[Choline][*seco*-Amx]	0.5	>10	0.05	>100	0.5	10
Na[*seco*-Amx]	>5	-	>5	-	>5	-
Amx	>5	1	>5	1	>5	1
[EMIM][*seco*-Pen]	>5	-	>5	-	>5	-
[C_2OHMIM][seo-Pen]	>5	-	>5	-	>5	-
[Choline][*seco*-Pen]	1.0	>5	>5	-	1.0	>5
[$P_{6,6,6,14}$][*seco*-Pen]	0.5	>10	0.5	>10	>5	-
[C_{16}Pyr]*seco*-[Pen]	0.5	>10	0.5	>10	0.05	>100
[TEA][*seco*-Pen]	>5	-	> 5	-	>5	-
K[*seco*-Pen]	>5	-	>5	-	>5	-
K[Pen]	>5	1	>5	1	>5	1

As expected, parent antibiotics (Amx and [K][Pen]), as well as the sodium and potassium salts of their ammonia hydrolysates, were found to be inactive against all tested resistant bacteria strains. For *E. coli*, five OSILs that contained the [*seco*-Amx] anion and three that contained the [*seco*-Pen] anion showed increased activity (RDIC values between >5 and >100) against the parent antimicrobials. The highest activity (RDIC > 100) was recorded for [C_{16}Pyr][*seco*-Amx] against both resistant *E. coli* strains, while [$P_{6,6,6,14}$][*seco*-Amx] and [Choline][*seco*-Amx] were selective towards only one of the strains—CTX M9 and CTX M2, respectively.

These results were somewhat similar to those regarding ampicillin-based API-OSILs against resistant Gram-negative *E. coli* species that were previously described by us [16], where [C_{16}Pyr][Amp] and [$P_{6,6,6,14}$][Amp] showed the highest antimicrobial activities of all compounds tested. Therein we suggested that the drug delivery of the APIs is enhanced in some resistant *E. coli* strains by the lipophilic counter-ions through the permeation of the outer layer. This postulation is supported by results from other authors. More specifically, Vincent et al. [80] and Langgartner, J. et al. [81] similarly demonstrated that transport across biological membranes bearing a highly polar anionic framework can be facilitated if the APIs are paired with lipophilic ammonium ions that act as phase transfer agents. Additionally, Rogers et al. [82] recently demonstrated on synthetic membrane models that API-OSILs that contain lipophilic cations, preferably with established hydrogen bonds, exhibit an increased membrane transport as compared to API-OSILs with weaker electrostatic interactions or even traditional halide or metal salts. Finally, it is important to note that both hydrophobic and hydrophilic ionic liquids have been recently studied as penetration enhancers [16,20,83,84]. Various nanocarriers can serve as antimicrobial enhancers because they incorporate into lipid membranes or cell walls, leading to membrane or wall disruptions in both Gram-positive and Gram-negative bacteria strains [77,85–91], therefore increasing the drug's efficiency.

The activity of prepared OSILs against the Gram-positive MRSA ATCC 43,300 strain seems even more peculiar. IC_{50} values as low as 5 and 50 μM, respectively, were obtained with [C_{16}Pyr][*seco*-Amx] and [C_{16}Pyr][*seco*-Pen], which corresponded to RDICs higher than 1000 and 100, respectively. The only other measurable value was obtained with [Choline][*seco*-Pen] (RDIC > 5). In the former cases, the contribution from the [C_{16}Pyr] cation was unquestionable. Literature data [92] have shown that the MIC for [C_{16}Pyr][Cl] is five times higher in the methicillin-resistant than in methicillin-sensitive *S. aureus* strains, suggesting that the contribution of this cation to the antibacterial activity of these particular API-OSILs [92] is particularly significant as opposed to the parent antibiotics or their hydrolyzed analogues (see Table 3). In other words, the amplified activity of [C_{16}Pyr][*seco*-Amx] and [C_{16}Pyr][*seco*-Pen] can only be achieved due to a synergic effect of both ionic species. Such a strong influence of $[C_{16}Pyr]^+$ seems quite in contrast with the unspecific activity of so-called enhancers (such as $[C_2OHMIM]^+$ for sensitive *S. aureus* above) and prompt us to consider it as a β-lactam antibiotic potentiator [63,72]. Similar results from the literature have shown that berberine in the presence of ampicillin and oxacillin markedly lowers their MICs against MRSA despite berberine alone exhibiting no bactericidal activity. Later, it was shown that, in fact, berberine affected MRSA biofilm development and the dissemination of biofilm-associated infections [93,94]. This effect is most likely related with the formation of salt bridges at an allosteric site of the PBP2a in MRSA ATCC 43300 [92]. Curiously, similar interactions were also found for another alkylpyridinium compound, namely the antibiotic ceftaroline [95]. Thus, an analogous ionic allosteric effect at PBP2a may be occurring with the described [C_{16}Pyr]-based OSILs. Further studies, namely molecular dynamic and docking simulations, must be performed in the future to confirm this assumption. Regardless of the mechanism, the pronounced increase in RDIC values of the prepared API-OSILs, particularly against resistant species, seems very interesting for a potential drug combination strategy. In spite of some controversy, the combination of antimicrobials with non-active compounds may provide a quite promising strategy to address the widespread emergence of antibiotic-resistant bacteria strains [96–98].

3. Conclusions

The present work highlights that organic salts and ionic liquids that contain ammonia hydrolysates of amoxicillin and penicillin G (*seco*-Amx and *seco*-Pen), in particular [C_2OHMIM][*seco*-Amx], [C_2OHMIM][*seco*-Pen], [TEA][*seco*-Pen], [C_{16}Pyr][*seco*-Amx], abd [Choline][*seco*-Amx], [C_{16}Pyr][*seco*-Pen] display a very strong antibacterial effect on sensitive and resistant *E. coli* and MRSA strains, respectively. The gathered data suggest that the adequate ionic pairing of such hydrolyzed antimicrobials with an ion-pair effect is vital to enhance or promote antibiotic activity, with possible alterations in their mechanism of action according to the selected counter-ion. In particular, the hydrophobic combination [C_{16}Pyr][*seco*-Amx] demonstrated the highest efficiency towards resistant bacteria strains, with particular emphasis to MRSA ATCC 43300. The combination of [C_{16}Pyr] with [*seco*-Pen] was also very effective against the latter. These results are clearly promising and point towards a beneficial effect on the drug delivery of the modified APIs when combined with hydrophobic organic cations. In this way, the antimicrobial resistance to these standard β-lactam antibiotics can be drastically reduced in vitro.

Our results also show that future developments of novel APIs-OSILs must not only focus on the toxicity and hydrophobicity of the counter ion—they must also look at the outcome. More specifically, [$P_{6,6,6,14}$][*seco*-Amp] follows the trend of [C_{16}Pyr][Amp] on *E. coli* resistant strains, which suggests that there may be other factors at stake to be considered.

The virtually limitless number of ionic pairs that can be assembled as API-OSILs makes this area of research particularly interesting and potentially thriving. In addition, the straightforward synthetic procedure adds virtually no barriers to its future industrial up-scaling and will thus eventually lead to an effective combination therapy model to tackle the ever-emerging bacterial resistance towards antibiotics.

4. Experimental Section

4.1. Synthesis

Commercially available reagents were purchased from Aldrich, BDH—the Frilabo and Solchemar laboratory reagents were used as received. The solvents were from Valente & Ribeiro and distilled before use. Whenever necessary, the solvents were dried by standard procedures, distilled under nitrogen and stored over molecular sieves.

The basic anion-exchange resin Amberlite IRA-400-OH (ion-exchange capacity 1.4 eq.mL^{-1}) and Amberlyst A-26 resins were purchased from Supelco. ^{1}H and ^{13}C-NMR spectra in $(CD_3)_2SO$, CD_3OD or D_2O (from Euriso-Top) were recorded on a Bruker AMX400 spectrometer at room temperature unless specified otherwise. To perform NMR, 5 mm borosilicate tubes were used, and the sample concentration was, approximately, 7 mg/mL for ^{1}H-NMR and 37 mg/mL for ^{13}C-NMR. Chemical shifts are reported downfield in parts per million (ppm).

ESI-MS were acquired with an API-ION TRAP(PO03MS), ITQB, Oeiras, Portugal, operating in both positive- and negative ion modes and equipped with a Z-spray source. Source and desolvation temperatures were 80 and 100 °C, respectively. The ionic liquid solutions in methanol at concentrations ~10–4 mol dm^{-3} were introduced at a 10 μl min^{-1} flow rate. The capillary and the cone voltage were 2600 and 25 V, respectively. Nitrogen was used as a nebulization gas and argon was used as a collision gas. ESI-MS-MS were acquired by selecting the precursor ion with the quadrupole and then performing collisions with argon at energies from 3 to 20 eV in the hexapole.

IR spectra were measured on a Perkin Elmer 683 by using KBr sample disks. Optical rotations were recorded on a Perkin Elmer 241MC. The melting temperature (mT) was determined with a melting point apparatus (Stuart Scientific). The elemental analysis experiments were performed in a CHNS Series Thermo Finnigan-CE Instruments Flash EA 1112 under standard conditions (T combustion reactor 900 °C, T GC column furnace 65 °C, multiseparation SS GC column, He_2 flow 130 mL/min, O_2 flow 250 mL/min). The penicillin G potassium was transformed in penicillin G ammonium by the adaptation of the method of Salivar, C. J. et al. [63]. Figures 2–16 illustrate the chemical structures of all prepared OSILs based APIs.

4.1.1. Synthesis of seco-Pen-Based OSILs

The preparation of ammonium (2S,5R,6R)-3,3-dimethyl-7-oxo-6-(2-phenylacetamido)-4-thia-1-azabicyclo [3.2.0]heptane-2-carboxylate [NH_4][Pen] was done as indicated by Salivar et al method.

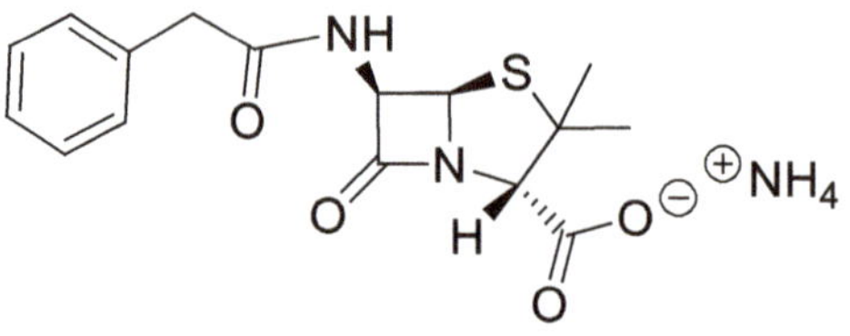

Figure 2. [NH_4][Pen].

The penicillin G potassium was transformed in penicillin G ammonium following the method of Salivar, C. J. et al. [63] before the reaction with hydroxide reactants.

Preparation of [NH_4][*seco*-Pen]

Potassium penicillin (1 g; 2.6 mmol) was dissolved in 15 mL of a 1.0 M aqueous ammonium solution. To the solution was added Amberlyst A-26 resin (5 eq.) that was previously stirred in a 2.0 M aqueous ammonium solution for 1 hour. The reaction mixture was stirred at room temperature for an additional 1 h. The resin was filtered off, and the was solvent evaporated to provide the desired product as a grey solid (0.828 g; 83%); ^{1}H NMR (400 MHz, D_2O) δ(ppm): 7.33–7.43 (m, 5H); 4.99 (d, *J* = 8 Hz, 1H); 4.81(s, 1H); 4.41 (d, *J* = 7.6 Hz, 1H); 3,69 (s, 2H); 3.50 (s, 1H); 1.56 (s, 3H); 1.25 (s, 3H); ^{13}C NMR (100 MHz, D_2O) δ(ppm): 177.55; 177.40; 176.65; 137.54; 132.14; 131.79; 130.22; 77.42; 67.24; 61.68; 60.93; 44.87; 29.56; 29.30.IR: υ = 3171; 2964; 1644; 1571; 1494; 1454; 1381; 1188; 1130; 1073; 1032; 784; 692; 502. Elemental analysis calculated for $C_{16}H_{24}N_4O_4S{\cdot}0.8H_2O$: C 50.19; H 6.74; N 14.63. Found: C 50.11; H 6.67; N 14.97.

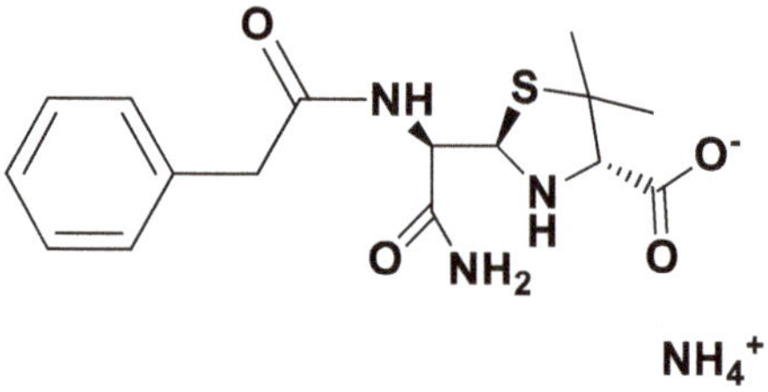

Figure 3. [NH_4][*seco*-Pen].

Preparation of [K][*seco*-Pen]

Potassium penicillin (0.137g; 0.36 mmol) was dissolved in a 1.0 M aqueous ammonium solution. The mixture was stirred at room temperature for 4 h. The solvent was evaporated to provide the desired product as a yellow solid (0.139 g; 97%); m.p. 193–195 °C; ^{1}H NMR (400 MHz, D_2O) δ(ppm): 7.34–7.44 (m, 5H); 4.98 (d, *J* = 7.6 Hz, 1H); 4.83 (t, *J* = 1.6 Hz, 1H); 4.39 (d, *J* = 8 Hz, 1H); 3.70 (s, 2H); 3.47 (s, 1H); 1.57(s, 3H); 1.25 (s, 3H); ^{13}C NMR (100 MHz, D_2O) δ(ppm): 177.82; 177.44; 176.79; 137.56; 131.77; 132.13; 130.19; 77.58; 67.42; 62.05; 61.20; 44.85; 29.49; 29.37; IR: υ = 3288; 2969; 2925; 1648; 1582; 1496; 1454; 1382; 1364; 1257; 1128; 877; 791; 694; 502. Elemental analysis calculated for $C_{16}H_{20}KN_4O_5S{\cdot}3H_2O$: C 44.22; H 5.80; N 9.67. Found: C 44.45; H 5.54; N 9.70.

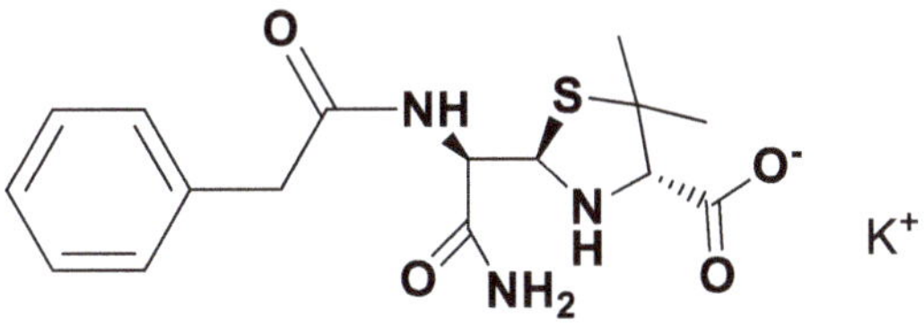

Figure 4. [K][*seco*-Pen].

Preparation of [Na][*seco*-Amx]

Amoxicillin (0.127 g; 0.3 mmol) was dissolved in a 1.0M aqueous ammonium solution. After 20 min, NaOH (0.012 g; 0.3 mmol) was added, and the mixture was stirred at room temperature for 4 h. The solvent was evaporated to provide the desired product as a yellow solid (0.127 g; 96%); m.p. 137–139 °C; ^{1}H NMR (400 MHz, D_2O) δ(ppm): 7.30 (d, *J* = 8 Hz, 2H); 6.89 (d, *J* = 8.4 Hz, 2H); 4.35 (d, *J* = 7.2 Hz, 1H); 3.29 (s, 1H); 1.38 (s, 3H); 1.17 (s, 3H); ^{13}C NMR (100 MHz, D_2O) δ(ppm): 177.77; 177.05; 176.60; 158.88; 131.63; 118.81; 77.61; 67.33; 61.91; 60.77; 60.23; 29.05; 28.85; IR: υ = 3277; 2969; 2919; 1648; 1575; 1510; 1433; 1383; 1322; 1245; 1173; 1127; 981; 865; 817; 780. Elemental analysis calculated for $C_{16}H_{21}N_4NaO_5S{\cdot}3H_2O$: C 41.92; H 5.94; N 12.22. Found: C 41.87; H 6.00; N 11.99.

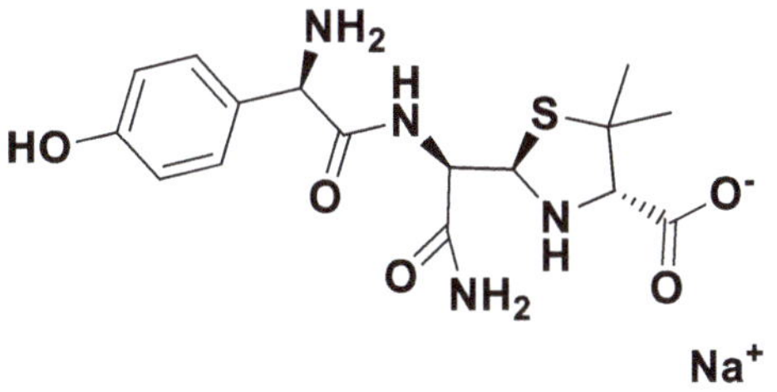

Figure 5. [Na][*seco*-Amx].

Preparation of [TEA][*seco*-Pen]

Tetraethylammonium bromide (0.420 g; 2.00 mmol) was dissolved in methanol and passed through Amberlite IRA-400-OH an ion-exchange column [18,99] (5 eq., flux rate 0.133 $mLmL^{-1}min^{-1}$ = 8 BVh^{-1}). Then, the tetraethylammonium hydroxide solution that was formed was slowly added to the ammonium penicillin G (0.751 g; 2.14 mmol) that was dissolved in the 1.0 M aqueous ammonium solution (50 $mgmL^{-1}$). The reaction mixture was stirred at room temperature for 1 h. After solvent evaporation, the residue was dissolved in a 20 mL solution (methanol/acetonitrile 1:9) [18,99] and left refrigerated overnight (4 °C) [18,99] to induce the precipitation of the excess of reagents. When the reagent crystals were filtered out, the solution was evaporated, and the rest was dried in a vacuum for 24 h to provide the desired product as a yellow viscous liquid (0.856 g; 90%). $[\alpha]_D^{25}$ = 104.0 ± 6.1 (c = 2 $mgmL^{-1}$ in methanol), ^{1}H-NMR (400.13 MHz, CD_3OD) δ = 7.31–7.22 (m, 5H), 5.46 (s, 1H), 4.17 (s, 1H) 3.70 (s, 1H), 3.63-3.58 (m, 2H), 3.51 (bs, 1H), 3.27–3.26 (m, 8H), 1.63 (s, 3H), 1.55 (s, 3H), 1.30–1.24 (m, 12H); ^{13}C-NMR (100.62 MHz, CD_3OD) δ = 174.72, 174.37, 174.15, 140.88, 136.73, 130.31, 130.25, 129.63, 127.96, 75.73, 66.68, 60.19, 59.27, 53.29, 46.65, 43.83, 29.41, 28.67, 28.38, 27.66, 7.64 ppm; IR (KBr): υ = 3420, 2981, 2924, 2862, 1840, 1736, 1721, 1648, 1560, 1543, 1490, 1459, 1432, 1396, 1367, 1173, 1130, 1053, 1027, 1001, 785, 734, 696, 619, 539 cm^{-1}; (ESI^+) *m/z* calculated for $C_8H_{20}N^+$: 130.1 found 130.0; (ESI^-) *m/z* calculated for $C_{16}H_{20}N_3O_4S^-$ 350.1 found 349.8.

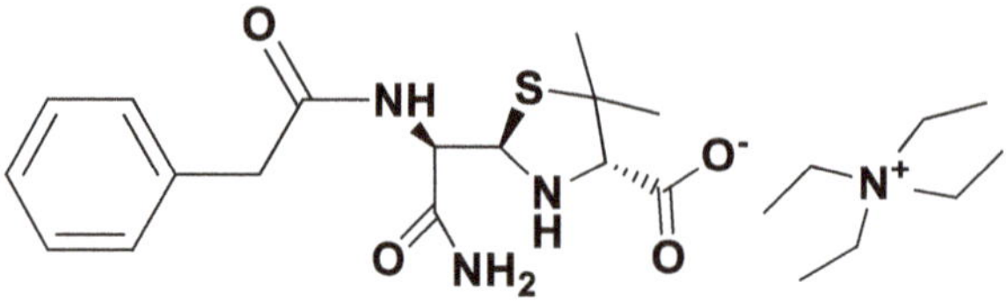

Figure 6. [TEA][*seco*-Pen].

Preparation of [$P_{6,6,6,14}$][*seco*-Pen]

Trihexyl(tetradecyl)phosphonium chloride (1.000 g; 1.92 mmol) was dissolved in methanol and passed through an Amberlite IRA-400-OH [18,99] ion-exchange column (5 eq., flux rate 0.133 $mLmL^{-1}min^{-1}$ = 8 BVh^{-1}). Then, the trihexyl(tetradecyl)phosphonium hydroxide solution that was formed was slowly added to ammonium penicillin G (0.853 g; 2.43 mmol) and dissolved in a 1.0 M aqueous ammonium solution (50 $mgmL^{-1}$). The mixture was stirred at room temperature for 1 h. After solvent evaporation, the residue was dissolved in a 20 mL solution (methanol/acetonitrile 1:9) [18,99] and left refrigerated overnight (4 °C) [14] to induce the precipitation of the excess reagent. When the reagent crystals were filtered out, the solution was evaporated, and the rest was dried in a vacuum for 24 h to provide the desired product as a yellow viscous liquid (1.560 g; 97%). $[\alpha]_D^{25}$ = 67.7 ± 3.0 (c = 2 $mgmL^{-1}$ in methanol); ^{1}H-NMR (400.13 MHz, CD_3OD) δ = 7.34–7.22 (m, 5H), 4.97 (1H, d, *J* = 6.7 Hz), 4.34 (dd, 1H, *J* = 6.7 Hz), 3.60 (d, 2H, *J* = 8.2 Hz), 3.50 (s, 1H), 2.20 (m, 8H), 1.56–1.25 (m, 54H), 0.96–0.88 (m, 12H) ppm; ^{13}C-NMR (100.62 MHz, CD_3OD) δ = 175.30, 174.83, 173.92, 136.76, 130.37, 129.59, 127.87, 76.81, 66.78, 60.14, 54.86, 43.68, 33.11, 32.19, 31.92, 31.84, 30.80, 30.51, 30.45, 29.91, 27.88, 23.77, 23.49, 22.36, 19.53, 19.05, 14.51, 14.38 ppm; IR (KBr): ν = 3308, 3028, 2951, 2923, 2853, 1737, 1669, 1607, 1536, 1496, 1456, 1418, 1379, 1262, 1201, 1113, 1031, 986, 860, 810, 761, 722, 694, 617, 454, 439, 424 cm^{-1}; (ESI^+) *m/z* calculated for $C_{32}H_{68}P^+$: 483.4 found 483.8; (ESI^-) *m/z* calculated for $C_{16}H_{20}N_3O_4S^-$ 350.1, found 349.9.

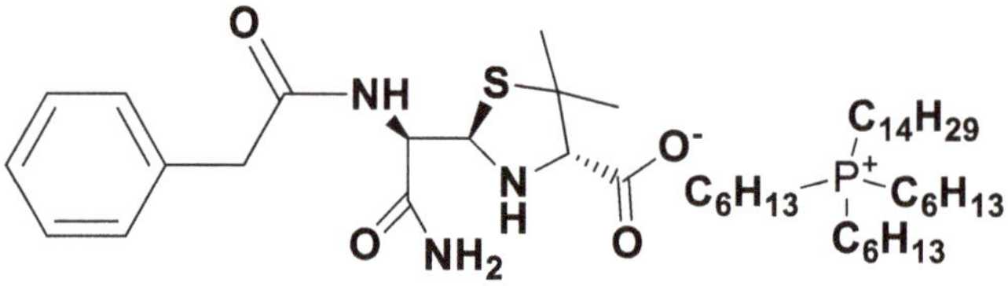

Figure 7. [$P_{6,6,6,14}$][*seco*-Pen].

Preparation of [C_{16}Pyr][*seco*-Pen]

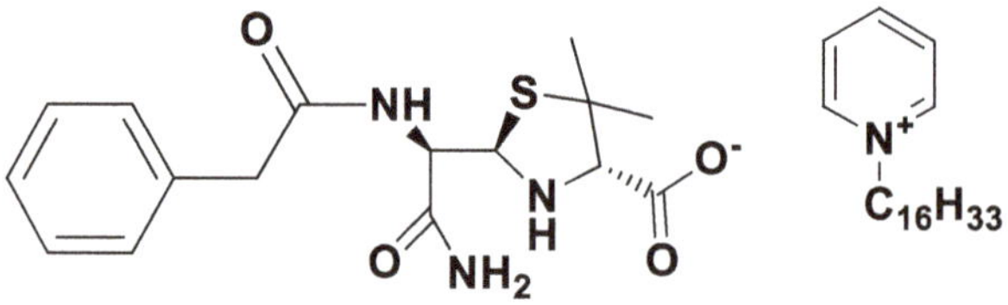

Figure 8. [C_{16}Pyr][*seco*-Pen].

Procedure I

Cetylpyridinium chloride (0.822 g; 2.30 mmol) was dissolved in methanol and passed through an Amberlite IRA-400-OH ion-exchange column [18,99] (5 eq., flux rate 0.133 $mLmL^{-1}min^{-1}$).

Then, the cetylpyridinium hydroxide solution that was formed was slowly added to ammonium penicillin G (0.973 g; 2.77 mmol) that was dissolved in a 1.0 M aqueous ammonium solution (50 $mgmL^{-1}$). The mixture was stirred at room temperature for 1 h. After solvent evaporation, the residue was dissolved in a 20 mL solution (methanol/acetonitrile 1:9) [18,99] and left refrigerated overnight (4 °C) [18,99] to induce the precipitation of the excess reagent. When the reagent crystals were filtered out, the solution was evaporated, and the rest was dried in a vacuum for 24 h to provide the desired product as a yellow solid (1.332 g; 89%). m.p. 76–78 °C; $[\alpha]_D^{25}$ = 47.3 ± 3.6 (c = 2 $mgmL^{-1}$ in methanol); ^{1}H-NMR (400.13 MHz, CD_3OD) δ = 9.01 (d, 2H, *J* = 5.7 Hz), 8.59 (t, 1H, *J* = 7.8 Hz), 8.12 (t, 2H, *J* = 6.8 Hz), 7.33–7.21 (m, 5H), 4.95 (d, 1H, *J* = 7.1 Hz), 4.63 (t, 2H, *J* = 7.5Hz,), 4.35 (d, 1H, *J* = 7.0 Hz), 3.60 (2H, d, *J* = 7.5 Hz), 3.50 (s, 1H), 1.56 (m, 3H), 1.42-1.09 (m, 31H), 0.90 (t, 3H, *J* = 6.6 Hz) ppm; ^{13}C-NMR (100.62 MHz, CD_3OD) δ = 175.16, 174.78, 173.98, 146.87, 146.00, 136.73, 130.40, 129.80, 129.60, 127.93, 76.34, 66.64, 63,15, 60.14, 43.67, 33.12, 32.55, 30.81, 30.67, 30.17, 27.86, 27.24, 23.78, 14.49 ppm; IR (KBr): δ = 3041, 3059, 2914, 2848, 1739, 1658, 167, 1601, 1542, 1528, 1508, 1487, 1472, 1397, 1368, 1322, 1270, 1209, 1177, 1128, 1078, 1032, 987, 960, 926, 818, 777, 716, 686, 619, 574, 475 cm^{-1}; (ESI^+) *m/z* calculated for $C_{21}H_{38}N^+$: 304.3 found 304.2; (ESI^-) *m/z* calculated for $C_{16}H_{20}N_3O_4S^-$ 350.1, found 349.9.

Procedure II

Cetylpyridinium chloride (0.145g; 0.43 mmol) was dissolved in methanol and was added Amberlyst A-26 (3 eq.) The mixture was stirred for 1 h at room temperature. Then, the cetylpyridinium hydroxide solution that was formed was slowly added to [NH_4][*seco*-Pen] (0.150 g; 0.43 mmol) that was dissolved in a 2.0 M aqueous ammonium solution, and the mixture was stirred at room temperature for 1 h. The solvent was evaporated to p rovide the desired product as a white solid (0.263 g; 94%).

Preparation of [Choline][*seco*-Pen]

(2-hydroxyethyl)trimethylammonium chloride (0.277 g; 1.99 mmol) was dissolved in methanol and passed through an Amberlite IRA-400-OH ion-exchange column [18,99] (5 eq., flux rate 0.133 $mLmL^{-1}min^{-1}$ = 8 BVh^{-1}). Then, the hydroxide solution that was formed was slowly added to ammonium penicillin G (0.848 g; 2.41 mmol) that was dissolved in a 1.0 M aqueous ammonium solution (50 $mgmL^{-1}$). The mixture was stirred at room temperature for 1 h. After solvent evaporation, the residue was dissolved in a 20 mL solution (methanol/acetonitrile 1:9) [18,99] and left refrigerated overnight (4 °C) [18,99] to induce the precipitation of the excess reagent. When the reagent crystals were filtered out, the solution was evaporated, and the rest was dried in a vacuum for 24 h to provide the desired product as a yellow solid (0.856 g; 95%). m.p. 69–71 °C; $[\alpha]_D^{25}$ = 47.3 ± 3.6 (c = 2 $mgmL^{-1}$ in methanol); ^{1}H-NMR (400.13 MHz, CD_3OD) δ = 7.33–7.29 (m, 5H), 4.95 (d, 1H, J_1 = 7.0 Hz), 4.35 (d, 1H, J_1 = 7.0 Hz), 4.02–3.98 (m, 2H), 3.66–3.56 (m, 2H), 3.50-3.47 (m, 3H), 3.20 (s, 9H) 1.56 (s, 3H), 1.25 (s, 3H) ppm; ^{13}C-NMR (100.62 MHz, CD_3OD) δ = 174.72, 174.37, 174.15, 140.88, 136.73, 130.31, 130.25, 129.63, 127.96, 75.73, 69.06, 66.67, 60.20, 60.04, 59.53, 57.10, 46.65, 27.78, 27.47 ppm; IR (KBr): υ = 3468, 3074, 2966, 1652, 1496, 1479, 1461, 1396, 1356, 1279, 1204, 1131, 1083, 1054, 1010, 955, 887, 867, 796, 734, 702, 670, 620, 540, 476 cm^{-1}; (ESI^+) *m/z* calculated for $C_5H_{14}NO^+$: 104.1, found 104.1; (ESI^-) *m/z* calculated for $C_{16}H_{20}N_3O_4S^-$ 350.1, found 349.9.

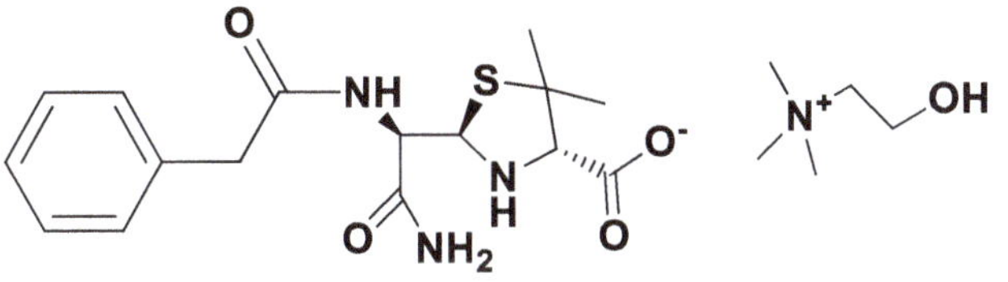

Figure 9. [Choline][*seco*-Pen].

Preparation of [EMIM][*seco*-Pen]

1-ethyl-3-methylimidazolium bromide (0.578 g; 3.03 mmol) was dissolved in methanol and passed through an Amberlite IRA-400-OH ion-exchange column [18,99] (5 eq., flux rate 0.133 $mLmL^{-1}min^{-1}$ = 8 BVh^{-1}). Then, the hydroxide solution that was formed was slowly added to ammonium penicillin G (1.11 g; 3.16 mmol) that was dissolved in an aqueous 1.0 M ammonium solution (50 $mgmL^{-1}$). The mixture was stirred at room temperature for 1 h. After solvent evaporation, the residue was dissolved in a 20 mL solution (methanol/acetonitrile 1:9) [18,99] and left refrigerated overnight (4 °C) [18,99] to induce the precipitation of the excess reagent. When the reagent crystals were filtered out, the solution was evaporated, and the rest was dried in a vacuum for 24 h to provide the desired product as a colorless viscous liquid (1.136 g; 81%). $[\alpha]_D^{25}$ = 89.0 ± 7.0 (c = 2 $mgmL^{-1}$ in methanol); ^{1}H-NMR (400.13 MHz, CD_3OD) δ = 8.99 (bs, 1H), 7.63 (s, 1H) 7.53 (s, 1H), 7.34–7.20 (m, 5H), 4.95 (d, 1H, *J* = 7.0 Hz), 4.35 (d, 1H, *J* = 7.0 Hz), 4.24 (q, 2H, *J* = 7.3 Hz), 3.92 (s, 3H), 3.60 (d, 2H, *J* = 7.2 Hz), 3.50 (s, 1H), 1,60–1.50 (m, 6H), 1.25 (s, 3H) ppm; ^{13}C-NMR (100.62 MHz, CD_3OD) δ = 175.06, 174.83, 173.99, 136.80, 130.65, 130.39, 129.60, 127.90, 124.94, 123.25, 76.51, 60.13, 60.18, 60.11, 59.54, 46.02, 43.66, 36.51, 27.81, 27.65, 15.61 pm; IR (KBr): υ = 3468, 3368, 2970, 1660, 1540, 1501, 1456, 1395, 1456, 1395, 1354, 1300, 1258, 1169, 1131, 1301, 965, 918, 862, 828, 729, 700, 651, 620, 545 cm^{-1}; (ESI^+) *m/z* calculated for $C_6H_{11}N_2^+$: 111.1, found 111.0; (ESI^-) *m/z* calculated for $C_{16}H_{20}N_3O_4S^-$ 350.1, found 349.9.

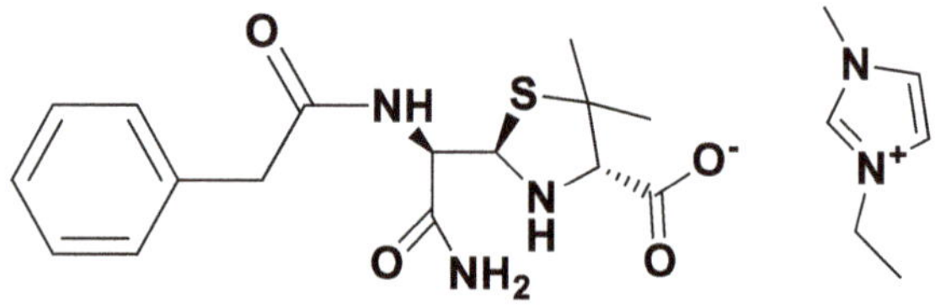

Figure 10. [EMIM][*seco*-Pen].

Preparation of [C_2OHMIM][*seco*-Pen]

3-(2-hydroxyethyl)-1-methylimidazolium chloride (0.328 g; 2.03 mmol) was dissolved in methanol and passed through an Amberlite IRA-400(OH) ion-exchange column [14,41] (5 eq., flux rate 0.133 $mLmL^{-1}min^{-1}$ = 8 BVh^{-1}). Then, the hydroxide solution that was formed was slowly added to ammonium penicillin G (0.754 g; 2.14 mmol) that was dissolved in a 1.0 M aqueous ammonium solution (50 $mgmL^{-1}$). The mixture was stirred at room temperature for 1 h. After solvent evaporation, the residue was dissolved in a 20 mL solution (methanol/acetonitrile 1:9) [18,99] and left refrigerated overnight (4 °C) [18,99] to induce the precipitation of the excess reagent. When the reagent crystals were filtered out, the solution was evaporated, and the rest was dried in a vacuum for 24 h to provide the desired product as a yellow solid (0.799 g; 83%). m.p. 48–50 °C; $[\alpha]_D^{25}$ = 41.3 ± 6.0 (c = 2 $mgmL^{-1}$ in methanol); ^{1}H-NMR (400.13 MHz, CD_3OD) δ = 8.98, (s, 1H), 7.61 (s, 1H), 7.54 (s, 1H), 7.33–7.21 (m, 5H), 4.94 (d, 1H, *J* = 7.1 Hz), 4.36 (d, 1H, *J* = 7.1 Hz), 4.29 (t, 2H, *J* = 4.9 Hz), 3.92, (s, 3H), 3.86 (t, 2H, *J* = 4.9 Hz), 3.59 (d, 2H, *J* = 7.1 Hz), 3.50 (s, 1H), 1.55 (s, 3H), 1.24 (s, 3H) ppm; ^{13}C-NMR (100.62 MHz, CD_3OD) δ = 175.04, 174.82, 174.00, 136.78, 130.63, 130.51, 130.40, 129.61, 129.85, 129.61, 127.91, 124.68, 124.00, 76.50, 66.76, 61.10, 60.12, 59.53, 53.26, 43.65, 36.45, 27.84 ppm; IR (KBr): υ = 3418, 2965, 2931, 2108, 1644, 1585, 1499, 1455, 1398, 1356, 1260, 1167, 1127, 1076, 1034, 879, 798, 734, 704, 668, 619, 464, 445, 432, 424 cm^{-1}; (ESI^+) *m/z* calculated for $C_6H_{11}N_2O^+$: 127.2, found 127.0; (ESI^-) *m/z* calculated for $C_{16}H_{20}N_3O_4S^-$ 350.1, found 349.9.

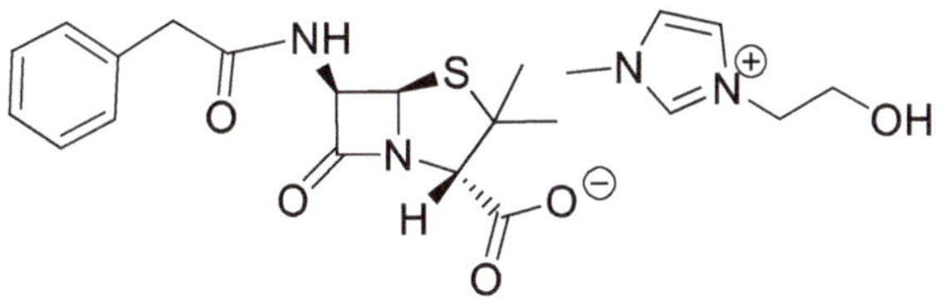

Figure 11. [C_2OHMIM][*seco*-Pen].

4.1.2. Synthesis of seco-Amx-Based OSILs

Preparation of [EMIM][*seco*-Amx]

1-ethyl-3-methylimidazolium chloride (0.385 g; 2.01 mmol) was dissolved in methanol and passed through an Amberlite IRA-400-OH ion-exchange column [18,99] (5 eq., flux rate 0.133 $mLmL^{-1}min^{-1}$ = 8 BVh^{-1}). Then, the hydroxide solution that was formed was slowly added to amoxicillin trihydrate (0.930 g; 2.22 mmol) that was dissolved in an aqueous 1.0 M ammonium solution (50 $mgmL^{-1}$). The mixture was stirred at room temperature for 1 h. After solvent evaporation, the residue was dissolved in a 20 mL solution (methanol/acetonitrile 1:9) [18,99] and left refrigerated overnight (4 °C) [18,99] to induce the precipitation of the excess reagent. When the reagent crystals were filtered out, the solution was evaporated, and the rest was dried in a vacuum for 24 h to provide the desired product as a yellow solid (0.768 g; 77%). m.p. 84–86 °C; $[\alpha]_D^{25}$ = 48.3 ± 5.0 (c = 2 $mgmL^{-1}$ in methanol); ^{1}H-NMR (400.13 MHz, CD_3OD) δ = 7.64 (s, 1H), 7.56 (s, 1H), 7.27 (d, 2H, *J* = 8.2 Hz), 6.74 (d, 2H, *J* = 8.4 Hz), 5.00 (d, 1H, J_1 = 5.9 Hz), 4.74 (s, 1H), 4.30 (d, 1H, J_1 = 5.9 Hz), 4.25 (1, 2H, *J* = 7.3 Hz), 3.77 (bs, 1H), 3.92 (s, 3H), 3.73, (bs, 1H), 3.43 (bs, 1H), 3.35 (s, 1H, s), 1.55–1.48 (m, 6H); 1.22 (s, 3H) ppm; ^{13}C-NMR (100.62 MHz, CD_3OD) δ = 175.57, 175.15, 174.84, 141.24, 129.83, 129.12, 128.50, 124.96, 123.31, 77.12, 66.67, 60.18, 60.11, 59.54, 46.03, 36.46, 27.78, 27.47, 15.63 ppm; IR (KBr): υ = 3461, 2921, 2852, 1706, 1688,1656, 1636, 1560, 1541, 1508, 1461, 1403, 1348, 1260, 1170, 1130, 673, 620, 474, 422 cm^{-1}; (ESI^+) *m/z* calculated for $C_6H_{11}N_2^+$: 111.1, found 111.0; (ESI^-) *m/z* calculated for $C_{16}H_{21}N_4O_5S^-$ 381.1, found 380.8.

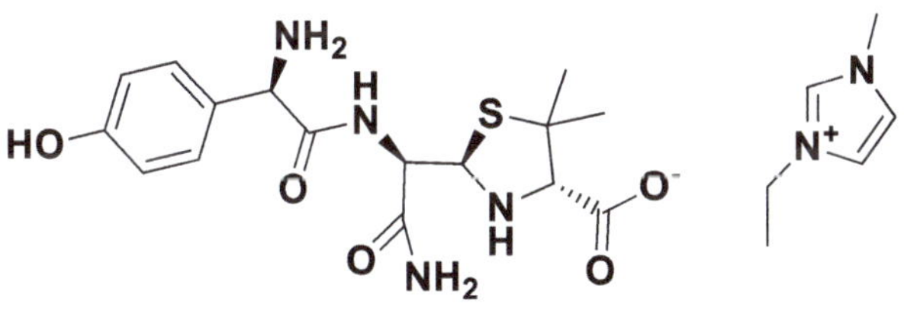

Figure 12. [EMIM][*seco*-Amx].

Preparation of [$P_{6,6,6,14}$][*seco*-Amx]

Trihexyl(tetradecyl)phosphonium chloride (1.042 g; 2.01 mmol) was dissolved in methanol and passed through an Amberlite IRA-400-OH ion-exchange column [18,99] (5 eq., flux rate 0.133 $mLmL^{-1}min^{-1}$ = 8 BVh^{-1}). Then, the trihexyl(tetradecyl)phosphonium hydroxide solution that was formed was slowly added to amoxicillin (0.988 g; 2.36 mmol) that was dissolved in a 1.0 M aqueous ammonium solution (50 $mgmL^{-1}$). The mixture was stirred at room temperature for 1 h. After solvent evaporation, the residue was dissolved in a 20 mL solution (methanol/acetonitrile 1:9) [18,99] and left refrigerated overnight (4 °C) [18,99] to induce the precipitation of the excess reagent. Then, the reagent crystals were filtered out, the solution was evaporated, and the rest was dried in a vacuum for 24 h to provide the desired product as a yellow viscous liquid (1.586 g; 92%). $[\alpha]_D^{25}$ = 22.0 ± 5.8 (c = 2 $mgmL^{-1}$ in methanol); (400.13 MHz, CD_3OD) δ = 7.28 (d, 2H, *J* = 8.4 Hz), 6.76 (d, 2H, *J* = 8.4 Hz), 5.00 (d, 1H,

J = 5.9 Hz), 4.53 (s, 1H), 4.31 (d, 1H, J = 5.9 Hz), 3.42 (s, 1H), 2.23–2.16 (m, 8H), 1.60–1.22 (m, 54H), 0.95–0.88 (m, 12H) ppm; ^{13}C-NMR (100.62 MHz, CD_3OD) δ = 175.35, 174.22, 142.01, 129.62, 128.92, 116.59, 77.12, 66.57, 60.17, 54.94, 43.76, 33.18, 32.27, 31.93, 31.56, 30.89, 30.58, 30.01, 27.96, 23.85, 23.57, 22.45, 19.62, 19.15, 14.58, 14.46 ppm; IR (KBr): ν = 3419, 3921, 2107.38, 1638, 1560, 1506, 1459, 1398, 1270, 1130, 1000, 668, 619, 570, 476, 456, 433, 412 cm^{-1}; (ESI$^+$) *m/z* calculated for $C_{32}H_{68}P^+$: 483.8 found 483.6; (ESI$^-$) *m/z* calculated for $C_{16}H_{21}N_4O_5S^-$ 381.1, found 381.0.

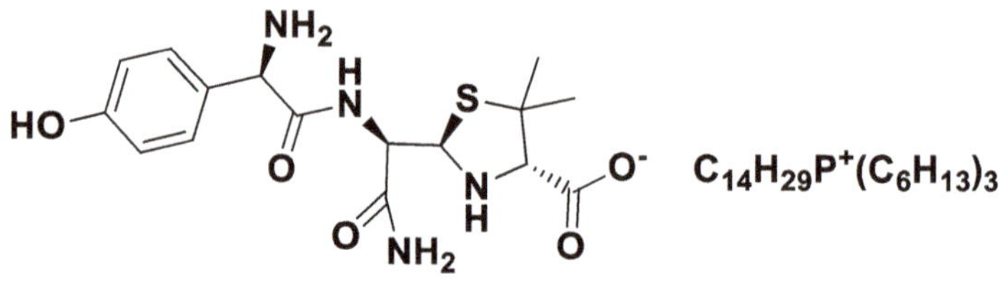

Figure 13. [$P_{6,6,6,14}$][*seco*-Amx].

Preparation of [C_{16}Pyr][*seco*-Amx]

Cetylpyridinium chloride (0.456 g; 1.28 mmol) was dissolved in methanol and passed through an Amberlite IRA-400-OH ion-exchange column [18,99] (5 eq., flux rate 0.133 mLmL^{-1}min^{-1} = 8 BVh^{-1}). Then, the cetylpyridinium hydroxide solution that was formed was slowly added to amoxicillin (0.587 g; 1.40 mmol) that was dissolved in a 1.0 M aqueous ammonium solution (50 mgmL^{-1}). The mixture was stirred at room temperature for 1 h. After solvent evaporation, the residue was dissolved in a 20 mL solution (methanol/acetonitrile 1:9) [18,99] and left refrigerated overnight (4 °C) [18,99] to induce the precipitation of the excess reagent. When the reagent crystals were filtered out, the solution was evaporated, and the rest was dried in a vacuum for 24 h to provide the desired product as a yellow solid (0.402 g; 52%). m.p. 96–98 °C; $[\alpha]_D^{25}$ = 77.0 ± 5.8 (c = 2 mgmL^{-1} in methanol); ^{1}H-NMR (400.13 MHz, CD_3OD) δ = 8.98 (d, 2H, J = 5.8 Hz), 8.58 (t, 1H, J = 7.74 Hz), 8.10 (t, 2H, J = 6.7 Hz), 7.26 (d, 2H, J = 8.5 Hz), 6.73 (d, 2H, J = 8.4 Hz), 5.01 (d, 1H, J = 6.0 Hz), 4.62 (t, 2H, J = 7.5 Hz), 4.46 (s, 1H), 4.30 (d, 1H, J = 6.0 Hz), 3.44 (s, 1H), 2.02, (t, 2H, J = 6.9 Hz), 1.48 (s, 3H), 1.38–1.26 (m, 28H), 1.22 (s, 3H) 0.90 (3H, t, J = 6.7 Hz) ppm; ^{13}C-NMR (100.62 MHz, CD_3OD) δ = 176.32, 175.60, 174.94, 158.32, 150.28, 146.87, 145.91, 132.86, 129.54, 116.55, 77.13, 66.61, 63,17, 60.14, 33.11, 32.53, 30.80, 30.67, 30.52, 30.16, 27.24, 23.77, 14.49 ppm; IR (KBr): υ = 3440, 2914, 2849, 1685, 1651, 1636, 1560, 1488, 1472, 1400, 1384, 1260, 1175, 1128, 847, 778, 720, 687, 621, 498, 476 cm^{-1}; (ESI$^+$) *m/z* calculated for $C_{21}H_{38}N^+$: 304.3 found 304.4; (ESI$^-$) *m/z* calculated for $C_{16}H_{21}N_4O_5S^-$ 381.1, found 380.9.

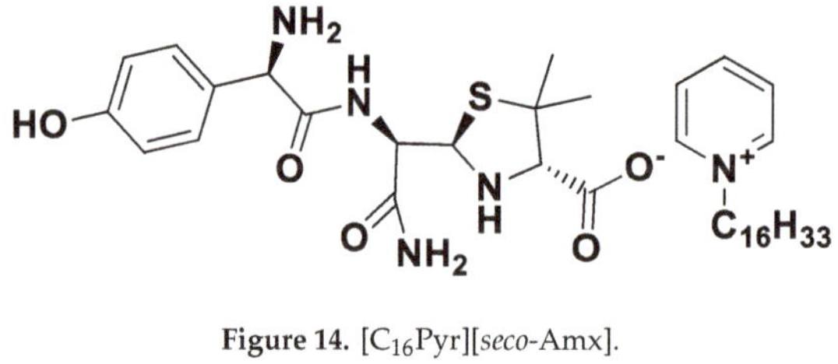

Figure 14. [C_{16}Pyr][*seco*-Amx].

Preparation [choline][*seco*-Amx]

(2-hydroxyethyl)trimethylammonium chloride (0.179 g; 1.28 mmol) was dissolved in methanol and passed through an Amberlite IRA-400-OH ion-exchange column [18,99] (5 eq., flux rate 0.133 mLmL^{-1}min^{-1} = 8 BVh^{-1}). Then, the hydroxide solution that was formed was slowly added to amoxicillin (0.587 g; 1.40 mmol) that was dissolved in a 1.0 M aqueous ammonium solution (50 mgmL^{-1}). The mixture was stirred at room temperature for 1 h. After solvent evaporation, the residue was dissolved in a 20 mL solution (methanol/acetonitrile 1:9) [18,99] and left refrigerated overnight (4 °C) [18,99] to induce the precipitation of the excess reagent. When the reagent crystals were filtered out, the solution was evaporated, and the rest was dried in a vacuum for 24 h to provide the desired product as a yellow solid (0.570 g; 93%). m.p. 143–144 °C; $[\alpha]_D^{25}$ = 104.0 ± 3.4 (c = 2 $mgmL^{-1}$ in methanol); ^{1}H-NMR (400.13 MHz, CD_3OD) δ = 7.36 (d, *J* = 8.1 Hz, 2H), 6.93 (d, *J* = 8.1 Hz, 2H), 5.06 (d, *J* = 6.8 Hz, 1H), 4.64 (s, 1H), 4.40 (d, *J* = 6.8 Hz, 1H), 4.13–4.04 (m, 2H), 3.53 (t, *J* = 4.6 Hz, 2H), 3.37 (s, 1H), 3.22 (s, 9H), 1.46 (s, 3H), 1.24 (s, 3H) ppm; ^{13}C-NMR (100.62 MHz, CD_3OD) δ = 176.14, 175.48, 174.46, 156.70, 131.47, 129.19, 116.54, 75.51, 67.98, 65.21, 59.78, 58.63, 58.46, 56.16, 54.48, 54.44, 54.40, 26.98, 26.79 ppm; IR (KBr): υ = 3300, 2964, 2927, 1673, 1594, 1513, 1435, 1389, 1251, 1118, 1130, 1087, 956, 837, 780 cm^{-1}.

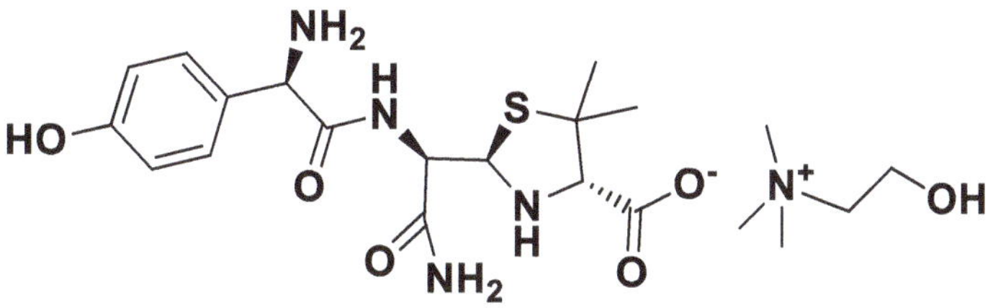

Figure 15. [choline][*seco*-Amx].

Preparation of [C_2OHMIM][*seco*-Amx]

3-(2-hydroxyethyl)-1-methylimidazolium chloride (0.456 g; 1.28 mmol) was dissolved in methanol and passed through an Amberlite IRA-400-OH ion-exchange column [18,99] (5 eq., flux rate 0.133 $mLmL^{-1}min^{-1}$ = 8 BVh^{-1}). Then, the hydroxide solution that was formed was slowly added to amoxicillin (0.525 g; 1.44 mmol) that was dissolved in a 1.0 M aqueous ammonium solution (50 $mgmL^{-1}$). The mixture was stirred at room temperature for 1 h. After solvent evaporation, the residue was dissolved in a 20 mL solution (methanol/acetonitrile 1:9) [18,99] and left refrigerated overnight (4 °C) [18,99] to induce the precipitation of the excess reagent. When the reagent crystals were filtered from the solution, the solution was evaporated, and the rest was dried in a vacuum for 24 h to provide the desired product as a yellow solid (0.359 g; 60%). m.p. 109–111 °C; $[\alpha]_D^{25}$ = 47.3 ± 3.6 (c = 2 $mgmL^{-1}$ in methanol); ^{1}H-NMR (400.13 MHz, CD_3OD) δ = 7.61 (s, 1H), 7.55 (s, 1H), 7.27 (d, 2H, *J* = 8.4 Hz), 6.74 (d, 2H, *J* = 8.4 Hz), 5.00 (d,1H, *J* = 6.0 Hz), 4.47 (s, 1H), 4.30–4.27 (m, 3H), 3.92 (s, 3H), 3.86 (t, 2H, *J* = 4.86 Hz), 3.43 (s, 1H), 1.48 (s, 3H), 1.22 (s, 3H) ppm; ^{13}C-NMR (100.62 MHz, CD_3OD) δ = 176.33, 175.62, 174.95, 158.34, 132.89, 129.54. 124.75,124.04, 116.56, 77.17 66.64, 61.11, 60.15, 59.83, 59.50, 53.81, 36.45, 27.78, 27.47 ppm; IR (KBr): υ = 3420, 2970, 2921, 1722, 1690, 1655, 1599, 1577, 1545, 1509, 1436, 1386, 1322, 1251, 1170, 1132, 1108, 1067, 877, 840, 820, 778, 652, 621, 535, 474 cm^{-1}; (ESI^+) *m/z* calculated for $C_6H_{11}N_2O^+$: 127.2, found 127.0; (ESI^-) *m/z* calculated for $C_{16}H_{21}N_4O_5S^-$ 381.1, found 380.9.

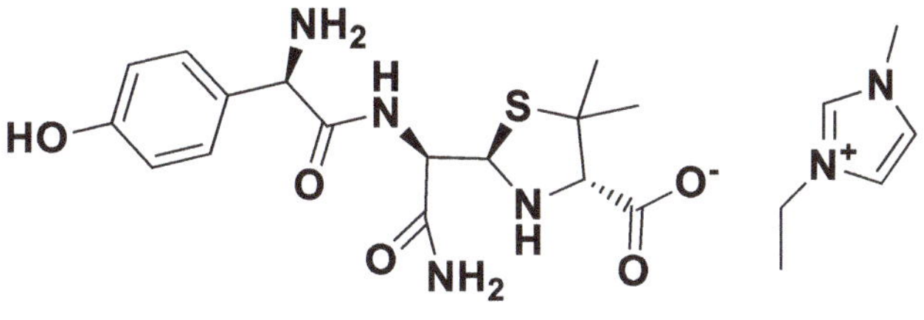

Figure 16. [C_2OHMIM][*seco*-Amx].

Author Contributions: The manuscript was written through contributions of all authors. Conceptualization, R.F.; Ž.P. and L.C.B.; methodology, R.F.; Ž.P. and L.C.B.; validation, R.F., M.M.S. and Ž.P.; investigation, R.F., D.S., A.R.D., V.D., L.P.; writing—original draft preparation, R.F., Ž.P. and M.M.S.; writing—review and editing, L.C.B.; visualization, C.P. and J.P.N. The first two authors contribute equally for the manuscript (share in equal parts as first author). All authors have read and agreed to the published version of the manuscript.

Funding: This research was funded by by Fundação para a Ciência e Tecnologia through projects (PTDC/QUI-QOR/32406/2017, PEst-C/LA0006/2013, PTDC/CTM/103664/2008) and PTDC/BTM-SAL/29786/2017, and one contract under Investigador FCT (L.C.B.). ZP thanks to Fundacão para a Ciência e a Tecnologia, MCTES, for the Norma transitória DL 57/2016 Program Contract. This work was also supported by the Associate Laboratory for Green Chemistry, which is financed by national funds from FCT/MEC (UID/QUI/50006/2013), UIDB/50006/2020 and co-financed by the ERDF under the PT2020 Partnership Agreement (POCI-01- 0145-FEDER-007265). The authors also thank Solchemar company for support.

Conflicts of Interest: The authors declare no conflict of interest.

References

1. Tseng, S.H.; Lee, C.M.; Lin, T.Y.; Chang, S.C.; Chang, F.Y. Emergence and spread of multi-drug resistant organisms: Think globally and act locally. *J. Microbiol. Immunol. Infect.* **2011**, *44*, 157–165. [CrossRef]
2. Livermore, D.M.; British Soc, A. Discovery research: The scientific challenge of finding new antibiotics. *J. Antimicrob. Chemother.* **2011**, *66*, 1941–1944. [CrossRef]
3. Martinez, J.L. General principles of antibiotic resistance in bacteria. *Drug Discov. Today Technol.* **2014**, *11*, 33–39. [CrossRef]
4. Coates, A.R.M.; Halls, G.; Hu, Y. Novel classes of antibiotics or more of the same? *Br. J. Pharmacol.* **2011**, *163*, 184–194. [CrossRef]
5. Laxminarayan, R.; Duse, A.; Wattal, C.; Zaidi, A.K.M.; Wertheim, H.F.L.; Sumpradit, N.; Vlieghe, E.; Hara, G.L.; Gould, I.M.; Goossens, H.; et al. Antibiotic resistance-the need for global solutions. *Lancet Infect. Dis.* **2013**, *13*, 1057–1098. [CrossRef]
6. Wright, G.D. Antibiotic Adjuvants: Rescuing Antibiotics from Resistance. *Trends Microbiol.* **2016**, *24*, 862–871. [CrossRef]
7. Laxminarayan, R.; Matsoso, P.; Pant, S.; Brower, C.; Rottingen, J.A.; Klugman, K.; Davies, S. Access to effective antimicrobials: A worldwide challenge. *Lancet* **2016**, *387*, 168–175. [CrossRef]
8. Amador, P.P.; Fernandes, R.M.; Prudencio, M.C.; Barreto, M.P.; Duarte, I.M. Antibiotic resistance in wastewater: Occurrence and fate of Enterobacteriaceae producers of Class A and Class C beta-lactamases. *J. Environ. Sci. Health Part A-Toxic/Hazard. Subst. Environ. Eng.* **2015**, *50*, 26–39. [CrossRef]
9. Vieira, M.; Pinheiro, C.; Fernandes, R.; Noronha, J.P.; Prudencio, C. Antimicrobial activity of quinoxaline 1,4-dioxide with 2-and 3-substituted derivatives. *Microbiol. Res.* **2014**, *169*, 287–293. [CrossRef]
10. Fernandes, R.; Amador, P.; Prudencio, C. beta-Lactams: Chemical structure, mode of action and mechanisms of resistance. *Rev. Med. Microbiol.* **2013**, *24*, 7–17. [CrossRef]
11. Woodford, N.; Turton, J.F.; Livermore, D.M. Multiresistant Gram-negative bacteria: The role of high-risk clones in the dissemination of antibiotic resistance. *Fems Microbiol. Rev.* **2011**, *35*, 736–755. [CrossRef]
12. Livermore, D.M. beta-Lactamases—the Threat Renews. *Curr. Protein Pept. Sci.* **2009**, *10*, 397–400. [CrossRef]
13. Coates, A.; Hu, Y.M.; Bax, R.; Page, C. The future challenges facing the development of new antimicrobial drugs. *Nat. Rev. Drug Discov.* **2002**, *1*, 895–910. [CrossRef]
14. Ferraz, R.; Noronha, J.; Murtinheira, F.; Nogueira, F.; Machado, M.; Prudencio, M.; Parapini, S.; D'Alessandro, S.; Teixeira, C.; Gomes, A.; et al. Primaquine-based ionic liquids as a novel class of antimalarial hits. *RSC Adv.* **2016**, *6*, 56134–56138. [CrossRef]
15. Ferraz, R.; Costa-Rodrigues, J.; Fernandes, M.H.; Santos, M.M.; Marrucho, I.M.; Rebelo, L.P.N.; Prudencio, C.; Noronha, J.P.; Petrovski, Z.; Branco, L.C. Antitumor Activity of Ionic Liquids Based on Ampicillin. *ChemMedChem* **2015**, *10*, 1480–1483. [CrossRef]
16. Ferraz, R.; Teixeira, V.; Rodrigues, D.; Fernandes, R.; Prudencio, C.; Noronha, J.P.; Petrovski, Z.; Branco, L.C. Antibacterial activity of Ionic Liquids based on ampicillin against resistant bacteria. *RSC Adv.* **2014**, *4*, 4301–4307. [CrossRef]

17. Florindo, C.; Araujo, J.M.M.; Alves, F.; Matos, C.; Ferraz, R.; Prudencio, C.; Noronha, J.P.; Petrovski, Z.; Branco, L.; Rebelo, L.P.N.; et al. Evaluation of solubility and partition properties of ampicillin-based ionic liquids. *Int. J. Pharm.* **2013**, *456*, 553–559. [CrossRef]
18. Ferraz, R.; Branco, L.C.; Marrucho, I.M.; Araujo, J.M.M.; Rebelo, L.P.N.; da Ponte, M.N.; Prudencio, C.; Noronha, J.P.; Petrovski, Z. Development of novel ionic liquids based on ampicillin. *MedChemComm* **2012**, *3*, 494–497. [CrossRef]
19. Ferraz, R.; Branco, L.C.; Prudencio, C.; Noronha, J.P.; Petrovski, Z. Ionic Liquids as Active Pharmaceutical Ingredients. *Chemmedchem* **2011**, *6*, 975–985. [CrossRef]
20. Egorova, K.S.; Gordeev, E.G.; Ananikov, V.P. Biological Activity of Ionic Liquids and Their Application in Pharmaceutics and Medicine. *Chem. Rev.* **2017**, *117*, 7131–7189. [CrossRef]
21. Cojocaru, O.A.; Bica, K.; Gurau, G.; Narita, A.; McCrary, P.D.; Shamshina, J.L.; Barber, P.S.; Rogers, R.D. Prodrug ionic liquids: Functionalizing neutral active pharmaceutical ingredients to take advantage of the ionic liquid form. *MedChemComm* **2013**, *4*, 559–563. [CrossRef]
22. Bica, K.; Rijksen, C.; Nieuwenhuyzen, M.; Rogers, R.D. In search of pure liquid salt forms of aspirin: Ionic liquid approaches with acetylsalicylic acid and salicylic acid. *Phys. Chem. Chem. Phys.* **2010**, *12*, 2011–2017. [CrossRef]
23. Dias, A.R.; Costa-Rodrigues, J.; Fernandes, M.H.; Ferraz, R.; Prudêncio, C. The Anticancer Potential of Ionic Liquids. *ChemMedChem* **2017**, *12*, 11–18. [CrossRef]
24. Teixeira, S.; Santos, M.M.; Ferraz, R.; Prudêncio, C.; Fernandes, M.H.; Costa-Rodrigues, J.; Branco, L.C. A Novel Approach for Bisphosphonates: Ionic Liquids and Organic Salts from Zoledronic Acid. *ChemMedChem* **2019**, *14*, 1767–1770. [CrossRef]
25. Santos, M.M.; Raposo, L.R.; Carrera, G.V.S.M.; Costa, A.; Dionísio, M.; Baptista, P.V.; Fernandes, A.R.; Branco, L.C. Ionic Liquids and salts from Ibuprofen as promising innovative formulations of an old drug. *ChemMedChem* **2019**, *14*, 907–911. [CrossRef]
26. Marrucho, I.M.; Branco, L.C.; Rebelo, L.P.N. Ionic Liquids in Pharmaceutical Applications. *Annu. Rev. Chem. Biomol. Eng.* **2014**, *5*, 527–546. [CrossRef]
27. Branco, L.C.; Carrera, G.V.S.M.; Aires-de-Sousa, J.; Martin, I.L.; Frade, R.; Afonso, C.A.M. Physico-Chemical Properties of Task-Specific Ionic Liquids, Ionic Liquids: Theory, Properties, New Approaches. In *Ionic Liquids: Theory, Properties, New Approaches*; Kokorin, P.A., Ed.; InTech: London, UK, 2011.
28. Hough, W.L.; Smiglak, M.; Rodriguez, H.; Swatloski, R.P.; Spear, S.K.; Daly, D.T.; Pernak, J.; Grisel, J.E.; Carliss, R.D.; Soutullo, M.D.; et al. The third evolution of ionic liquids: Active pharmaceutical ingredients. *New J. Chem.* **2007**, *31*, 1429–1436. [CrossRef]
29. Dean, P.M.; Turanjanin, J.; Yoshizawa-Fujita, M.; MacFarlane, D.R.; Scott, J.L. Exploring an Anti-Crystal Engineering Approach to the Preparation of Pharmaceutically Active Ionic Liquids. *Cryst. Growth Des.* **2009**, *9*, 1137–1145. [CrossRef]
30. Florindo, C.; Costa, A.; Matos, C.; Nunes, S.L.; Matias, A.N.; Duarte, C.M.M.; Rebelo, L.P.N.; Branco, L.C.; Marrucho, I.M. Novel organic salts based on fluoroquinolone drugs: Synthesis, bioavailability and toxicological profiles. *Int. J. Pharm.* **2014**, *469*, 179–189. [CrossRef]
31. Araujo, J.M.M.; Florindo, C.; Pereiro, A.B.; Vieira, N.S.M.; Matias, A.A.; Duarte, C.M.M.; Rebelo, L.P.N.; Marrucho, I.M. Cholinium-based ionic liquids with pharmaceutically active anions. *RSC Adv.* **2014**, *4*, 28126–28132. [CrossRef]
32. Carson, L.; Chau, P.K.W.; Earle, M.J.; Gilea, M.A.; Gilmore, B.F.; Gorman, S.P.; McCann, M.T.; Seddon, K.R. Antibiofilm activities of 1-alkyl-3-methylimidazolium chloride ionic liquids. *Green Chem.* **2009**, *11*, 492–497. [CrossRef]
33. Hough, W.L.; Rogers, R.D. Ionic liquids then and now: From solvents to materials to active pharmaceutical ingredients. *Bull. Chem. Soc. Jpn.* **2007**, *80*, 2262–2269. [CrossRef]
34. Demberelnyamba, D.; Kim, K.S.; Choi, S.J.; Park, S.Y.; Lee, H.; Kim, C.J.; Yoo, I.D. Synthesis and antimicrobial properties of imidazolium and pyrrolidinonium salts. *Bioorg. Med. Chem.* **2004**, *12*, 853–857. [CrossRef] [PubMed]
35. McCrary, P.D.; Beasley, P.A.; Gurau, G.; Narita, A.; Barber, P.S.; Cojocaru, O.A.; Rogers, R.D. Drug specific, tuning of an ionic liquid's hydrophilic-lipophilic balance to improve water solubility of poorly soluble active pharmaceutical ingredients. *New J. Chem.* **2013**, *37*, 2196–2202. [CrossRef]

36. Shamshina, J.L.; Kelley, S.P.; Gurau, G.; Rogers, R.D. Chemistry: Develop ionic liquid drugs. *Nature* **2015**, *528*, 188–189. [CrossRef]
37. Cherukuvada, S.; Nangia, A. Polymorphism in an API ionic liquid: Ethambutol dibenzoate trimorphs. *Crystengcomm* **2012**, *14*, 7840–7843. [CrossRef]
38. Yang, T.; Gao, G. Ionic Liquids in Pharmaceuticals. *Prog. Chem.* **2012**, *24*, 1928–1935.
39. Sekhon, B.S. Ionic liquids: Pharmaceutical and biotechnological applications. *Asian J. Pharm. Biol. Res.* **2011**, *1*, 395–411.
40. Stoimenovski, J.; MacFarlane, D.R.; Bica, K.; Rogers, R.D. Crystalline vs. Ionic Liquid Salt Forms of Active Pharmaceutical Ingredients: A Position Paper. *Pharm. Res.* **2010**, *27*, 521–526. [CrossRef]
41. Smiglak, M.; Pringle, J.M.; Lu, X.; Han, L.; Zhang, S.; Gao, H.; MacFarlane, D.R.; Rogers, R.D. Ionic liquids for energy, materials, and medicine. *Chem. Commun.* **2014**, *50*, 9228–9250. [CrossRef]
42. Ferraz, R.; Pinheiro, M.; Gomes, A.; Teixeira, C.; Prudêncio, C.; Reis, S.; Gomes, P. Effects of novel triple-stage antimalarial ionic liquids on lipid membrane models. *Bioorg. Med. Chem. Lett.* **2017**, *27*, 4190–4193. [CrossRef]
43. Brunel, F.; Lautard, C.; Garzino, F.; Giorgio, S.; Raimundo, J.M.; Bolla, J.M.; Camplo, M. Antibacterial activities of fluorescent nano assembled triphenylamine phosphonium ionic liquids. *Bioorg. Med. Chem. Lett.* **2016**, *26*, 3770–3773. [CrossRef]
44. Kontro, I.; Svedstrom, K.; Dusa, F.; Ahvenainen, P.; Ruokonen, S.K.; Witos, J.; Wiedmer, S.K. Effects of phosphonium-based ionic liquids on phospholipid membranes studied by small-angle X-ray scattering. *Chem. Phys. Lipids* **2016**, *201*, 59–66. [CrossRef]
45. Busetti, A.; Crawford, D.E.; Earle, M.J.; Gilea, M.A.; Gilmore, B.F.; Gorman, S.P.; Laverty, G.; Lowry, A.F.; McLaughlin, M.; Seddon, K.R. Antimicrobial and antibiofilm activities of 1-alkylquinolinium bromide ionic liquids. *Green Chem.* **2010**, *12*, 420–425. [CrossRef]
46. Iwai, N.; Nakayama, K.; Kitazume, T. Antibacterial activities of imidazolium, pyrrolidinium and piperidinium salts. *Bioorg. Med. Chem. Lett.* **2011**, *21*, 1728–1730. [CrossRef]
47. Cole, M.R.; Hobden, J.A.; Warner, I.M. Recycling Antibiotics into GUMBOS: A New Combination Strategy to Combat Multi-Drug-Resistant Bacteria. *Molecules* **2015**, *20*, 6466–6487. [CrossRef]
48. Bergamo, V.Z.; Donato, R.K.; Dalla Lana, D.F.; Donato, K.J.Z.; Ortega, G.G.; Schrekker, H.S.; Fuentefria, A.M. Imidazolium salts as antifungal agents: Strong antibiofilm activity against multidrug-resistant Candida tropicalis isolates. *Lett. Appl. Microbiol.* **2015**, *60*, 66–71. [CrossRef]
49. Choi, S.Y.; Rodriguez, H.; Gunaratne, H.Q.N.; Puga, A.V.; Gilpin, D.; McGrath, S.; Vyle, J.S.; Tunney, M.M.; Rogers, R.D.; McNally, T. Dual functional ionic liquids as antimicrobials and plasticisers for medical grade PVCs. *RSC Adv.* **2014**, *4*, 8567–8581. [CrossRef]
50. Nancharaiah, Y.V.; Reddy, G.K.K.; Lalithamanasa, P.; Venugopalan, V.P. The ionic liquid 1-alkyl-3-methylimidazolium demonstrates comparable antimicrobial and antibiofilm behavior to a cationic surfactant. *Biofouling* **2012**, *28*, 1141–1149. [CrossRef]
51. Hu, D.Y.; Li, X.; Sreenivasan, P.K.; DeVizio, W. A Randomized, Double-Blind Clinical Study to Assess the Antimicrobial Effects of a Cetylpyridinium Chloride Mouth Rinse on Dental Plaque Bacteria. *Clin. Ther.* **2009**, *31*, 2540–2548. [CrossRef]
52. Coleman, D.; Spulak, M.; Teresa Garcia, M.; Gathergood, N. Antimicrobial toxicity studies of ionic liquids leading to a 'hit' MRSA selective antibacterial imidazolium salt. *Green Chem.* **2012**, *14*, 1350–1356. [CrossRef]
53. Bielaszewska, M.; Mellmann, A.; Zhang, W.; Koeck, R.; Fruth, A.; Bauwens, A.; Peters, G.; Karch, H. Characterisation of the Escherichia coli strain associated with an outbreak of haemolytic uraemic syndrome in Germany, 2011: A microbiological study. *Lancet Infect. Dis.* **2011**, *11*, 671–676. [CrossRef]
54. Pennington, H. Escherichia coli O104, Germany 2011. *Lancet Infect. Dis.* **2011**, *11*, 652–653. [CrossRef]
55. Brain, E.G.; Eglington, A.J.; Nayler, J.H.C.; Pearson, M.J.; Southgate, R. Oxidation of some 1,2-seco-penicillins. *J. Soc. Chem. Commun.* **1972**, *4*, 229–230. [CrossRef]
56. Gower, J.L.; Risbridger, G.D.; Redrup, M.J. Positive and negative-ion fast atom bombardment mass-spectra of some penicilloic acids. *J. Antibiot.* **1984**, *37*, 33–43. [CrossRef]
57. Hart, K.M.; Reck, M.; Bowman, G.R.; Wencewicz, T.A. Tabtoxinine-beta-lactam is a "stealth" beta-lactam antibiotic that evades beta-lactamase-mediated antibiotic resistance. *MedChemComm* **2016**, *7*, 118–127. [CrossRef]

58. Paula, M.V.; Barros, A.L.; Wanderley, K.A.; de Sa, G.F.; Eberlin, M.; Soares, T.A.; Alves, S. Metal Organic Frameworks for Selective Degradation of Amoxicillin in Biomedical Wastes. *J. Braz. Chem. Soc.* **2018**, *29*, 2127–2136. [CrossRef]
59. Davis, A.M.; Layland, N.J.; Page, M.I.; Martin, F.; Oferrall, R.M. Thiazolidine ring-opening in penicillin derivatives .2. Enamine formation. *J. Chem. Soc. Perkin Trans.* **1991**, *2*, 1225–1229. [CrossRef]
60. Styring, P.; Chong, S.S.F. Stereoselective synthesis of a thiazolane amide using molecular recognition in the triazolyl-activated ester intermediate. *Tetrahedron Lett.* **2006**, *47*, 1737–1740. [CrossRef]
61. Hamiltonmiller, J.M.; Richards, E.; Abraham, E.P. Changes in proton-magnetic-resonance spectra during aminolysis and enzymic hydrolysis of cephalosporins. *Biochem. J.* **1970**, *116*, 385–395.
62. Salivar, C.J.; Grenfell, T.C.; Brown, E.V. Studies on the naturally occurring penicillins. 2. Precipitation of crystalline ammonium penicillins. *J. Biol. Chem.* **1948**, *176*, 977–981.
63. NCCLS. *Methods for Dilution Antimicrobial Susceptibility Tests for Bacteria That Grow Aerobically: Approved Standard*; NCCLS: Wayne, PA, USA, 2003.
64. Cole, M.R.; Li, M.; El-Zahab, B.; Janes, M.E.; Hayes, D.; Warner, I.M. Design, Synthesis, and Biological Evaluation of beta-Lactam Antibiotic-Based Imidazolium- and Pyridinium-Type Ionic Liquids. *Chem. Biol. Drug Des.* **2011**, *78*, 33–41. [CrossRef]
65. Drawz, S.M.; Bonomo, R.A. Three Decades of β-Lactamase Inhibitors. *Clin. Microbiol. Rev.* **2010**, *23*, 160–201. [CrossRef]
66. Zapun, A.; Contreras-Martel, C.; Vernet, T. Penicillin-binding proteins and beta-lactam resistance. *Fems Microbiol. Rev.* **2008**, *32*, 361–385. [CrossRef]
67. Mainardi, J.L.; Fourgeaud, M.; Hugonnet, J.E.; Dubost, L.; Brouard, J.P.; Ouazzani, J.; Rice, L.B.; Gutmann, L.; Arthur, M. A novel peptidoglycan cross-linking enzyme for a beta-lactam-resistant transpeptidation pathway. *J. Biol. Chem.* **2005**, *280*, 38146–38152. [CrossRef]
68. Edoo, Z.; Arthur, M.; Hugonnet, J.E. Reversible inactivation of a peptidoglycan transpeptidase by a beta-lactam antibiotic mediated by beta-lactam-ring recyclization in the enzyme active site. *Sci. Rep.* **2017**, *7*, 1–8. [CrossRef]
69. Trapp, S.; Franco, A.; Mackay, D. Activity-Based Concept for Transport and Partitioning of Ionizing Organics. *Environ. Sci. Technol.* **2010**, *44*, 6123–6129. [CrossRef]
70. Shamshina, J.L.; Barber, P.S.; Rogers, R.D. Ionic liquids in drug delivery. *Expert Opin. Drug Deliv.* **2013**, *10*, 1367–1381. [CrossRef]
71. Shimizu, M.; Shiota, S.; Mizushima, T.; Ito, H.; Hatano, T.; Yoshida, T.; Tsuchiya, T. Marked potentiation of activity of beta-lactams against methicillin-resistant Staphylococcus aureus by corilagin. *Antimicrob. Agents Chemother.* **2001**, *45*, 3198–3201. [CrossRef]
72. Hu, Z.X.; Sun, W.G.; Li, Q.; Li, X.N.; Zhu, H.C.; Huang, J.F.; Liu, J.J.; Wang, J.P.; Xue, Y.B.; Zhang, Y.H. Spiroaspertrione A, a Bridged Spirocyclic Meroterpenoid, as a Potent Potentiator of Oxacillin against Methicillin-Resistant Staphylococcus aureus from Aspergillus sp TJ23. *J. Org. Chem.* **2017**, *82*, 3125–3131.
73. Wang, H.; Gill, C.J.; Lee, S.H.; Mann, P.; Zuck, P.; Meredith, T.C.; Murgolo, N.; She, X.W.; Kales, S.; Liang, L.Z.; et al. Discovery of Wall Teichoic Acid Inhibitors as Potential Anti-MRSA beta-Lactam Combination Agents. *Chem. Biol.* **2013**, *20*, 272–284. [CrossRef]
74. Löbenberg, R.; Amidon, G.L. Modern bioavailability, bioequivalence and biopharmaceutics classification system. New scientific approaches to international regulatory standards. *Eur. J. Pharm. Biopharm.* **2000**, *50*, 3–12. [CrossRef]
75. Amidon, G.L.; Lennernäs, H.; Shah, V.P.; Crison, J.R. A Theoretical Basis for a Biopharmaceutic Drug Classification: The Correlation of in Vitro Drug Product Dissolution and in Vivo Bioavailability. *Pharm. Res.* **1995**, *12*, 413–420. [CrossRef]
76. Bakshi, P.S.; Gusain, R.; Dhawaria, M.; Suman, S.K.; Khatri, O.P. Antimicrobial and lubrication properties of 1-acetyl-3-hexylbenzotriazolium benzoate/sorbate ionic liquids. *RSC Adv.* **2016**, *6*, 46567–46572. [CrossRef]
77. Fang, Y.Y.; Liu, T.L.; Zou, Q.L.; Zhao, Y.X.; Wu, F.P. Cationic benzylidene cyclopentanone photosensitizers for selective photodynamic inactivation of bacteria over mammalian cells. *RSC Adv.* **2015**, *5*, 56067–56074. [CrossRef]
78. Mester, P.; Jehle, A.K.; Leeb, C.; Kalb, R.; Grunert, T.; Rossmanith, P. FTIR metabolomic fingerprint reveals different modes of action exerted by active pharmaceutical ingredient based ionic liquids (API-ILs) on Salmonella typhimurium. *RSC Adv.* **2016**, *6*, 32220–32227. [CrossRef]

79. Corte, L.; Tiecco, M.; Roscini, L.; De Vincenzi, S.; Colabella, C.; Germani, R.; Tascini, C.; Cardinali, G. FTIR Metabolomic Fingerprint Reveals Different Modes of Action Exerted by Structural Variants of N-Alkyltropinium Bromide Surfactants on Escherichia coli and Listeria innocua Cells. *PLoS ONE* **2015**, *10*, e0115275. [CrossRef]
80. Vincent, S.P.; Lehn, J.M.; Lazarte, J.; Nicolau, C. Transport of the highly charged myo-inositol hexakisphosphate molecule across the red blood cell membrane: A phase transfer and biological study. *Bioorg. Med. Chem.* **2002**, *10*, 2825–2834. [CrossRef]
81. Langgartner, J.; Lehn, N.; Gluck, T.; Herzig, H.; Kees, F. Comparison of the pharmacokinetics of piperacillin and sulbactam during intermittent and continuous intravenous infusion. *Chemotherapy* **2007**, *53*, 370–377. [CrossRef]
82. Zavgorodnya, O.; Shamshina, J.L.; Mittenthal, M.; McCrary, P.D.; Rachiero, G.P.; Titi, H.M.; Rogers, R.D. Polyethylene glycol derivatization of the non-active ion in active pharmaceutical ingredient ionic liquids enhances transdermal delivery. *New J. Chem.* **2017**, *41*, 1499–1508. [CrossRef]
83. Dobler, D.; Schmidts, T.; Zinecker, C.; Schlupp, P.; Schafer, J.; Runkel, F. Hydrophilic Ionic Liquids as Ingredients of Gel-Based Dermal Formulations. *AAPS PharmSciTech* **2016**, *17*, 923–931. [CrossRef]
84. De Almeida, T.S.; Julio, A.; Caparica, R.; Rosado, C.; Fernandes, A.S.; Saraiva, N.; Ribeiro, M.; Araujo, M.E.; Baby, A.R.; Costa, J.; et al. Ionic liquids as solubility/permeation enhancers for topical formulations: Skin permeation and cytotoxicity characterization. *Toxicol. Lett.* **2015**, *238*, S293. [CrossRef]
85. Jiang, H.; Xiong, M.M.; Bi, Q.Y.; Wang, Y.; Li, C. Self-enhanced targeted delivery of a cell wall- and membrane-active antibiotics, daptomycin, against staphylococcal pneumonia. *Acta Pharm. Sin. B* **2016**, *6*, 319–328. [CrossRef]
86. Khameneh, B.; Iranshahy, M.; Ghandadi, M.; Atashbeyk, D.G.; Bazzaz, B.S.F.; Iranshahi, M. Investigation of the antibacterial activity and efflux pump inhibitory effect of co-loaded piperine and gentamicin nanoliposomes in methicillin-resistant Staphylococcus aureus. *Drug Dev. Ind. Pharm.* **2015**, *41*, 989–994. [CrossRef]
87. Pinilla, C.M.B.; Brandelli, A. Antimicrobial activity of nanoliposomes co-encapsulating nisin and garlic extract against Gram-positive and Gram-negative bacteria in milk. *Innov. Food Sci. Emerg. Technol.* **2016**, *36*, 287–293. [CrossRef]
88. Ma, Y.F.; Wang, Z.; Zhao, W.; Lu, T.L.; Wang, R.T.; Mei, Q.B.; Chen, T. Enhanced bactericidal potency of nanoliposomes by modification of the fusion activity between liposomes and bacterium. *Int. J. Nanomed.* **2013**, *8*, 2351–2360. [CrossRef]
89. Rout, B.; Liu, C.H.; Wu, W.C. Enhancement of photodynamic inactivation against Pseudomonas aeruginosa by a nano-carrier approach. *Colloids Surf. B-Biointerfaces* **2016**, *140*, 472–480. [CrossRef]
90. Lopes, L.B.; Garcia, M.T.J.; Bentley, M. Chemical penetration enhancers. *Ther. Deliv.* **2015**, *6*, 1053–1061. [CrossRef]
91. Atashbeyk, D.G.; Khameneh, B.; Tafaghodi, M.; Bazzaz, B.S.F. Eradication of methicillin-resistant Staphylococcus aureus infection by nanoliposomes loaded with gentamicin and oleic acid. *Pharm. Biol.* **2014**, *52*, 1423–1428. [CrossRef]
92. Irizarry, L.; Merlin, T.; Rupp, J.; Griffith, J. Reduced susceptibility of methicillin-resistant Staphylococcus aureus to cetylpyridinium chloride and chlorhexidine. *Chemotherapy* **1996**, *42*, 248–252. [CrossRef] [PubMed]
93. Chu, M.; Zhang, M.B.; Liu, Y.C.; Kang, J.R.; Chu, Z.Y.; Yin, K.L.; Ding, L.Y.; Ding, R.; Xiao, R.X.; Yin, Y.N.; et al. Role of Berberine in the Treatment of Methicillin-Resistant Staphylococcus aureus Infections. *Sci. Rep.* **2016**, *6*, 24748. [CrossRef] [PubMed]
94. Yu, H.H.; Kim, K.J.; Cha, J.D.; Kim, H.K.; Lee, Y.E.; Choi, N.Y.; You, Y.O. Antimicrobial activity of berberine alone and in combination with ampicillin or oxacillin against methicillin-resistant Staphylococcus aureus. *J. Med. Food* **2005**, *8*, 454–461. [CrossRef]
95. Fishovitz, J.; Hermoso, J.A.; Chang, M.; Mobashery, S. Penicillin-binding Protein 2a of Methicillin-resistant Staphylococcus aureus. *Iubmb Life* **2014**, *66*, 572–577. [CrossRef]
96. Tyers, M.; Wright, G.D. Drug combinations: A strategy to extend the life of antibiotics in the 21 st century. *Nat. Rev. Microbiol.* **2019**, *17*, 141–155. [CrossRef]
97. Ocampo, P.S.; Lazar, V.; Papp, B.; Arnoldini, M.; zur Wiesch, P.A.; Busa-Fekete, R.; Fekete, G.; Pal, C.; Ackermann, M.; Bonhoeffer, S. Antagonism between Bacteriostatic and Bactericidal Antibiotics Is Prevalent. *Antimicrob. Agents Chemother.* **2014**, *58*, 4573–4582. [CrossRef]

98. Murray, C.W.; Rees, D.C. The rise of fragment-based drug discovery. *Nat. Chem.* **2009**, *1*, 187–192. [CrossRef]
99. Fukumoto, K.; Yoshizawa, M.; Ohno, H. Room Temperature Ionic Liquids from 20 Natural Amino Acids. *J. Am. Chem. Soc.* **2005**, *127*, 2398–2399. [CrossRef]

Article

Antimicrobial Activities of Highly Bioavailable Organic Salts and Ionic Liquids from Fluoroquinolones

Miguel M. Santos [1,*], Celso Alves [2], Joana Silva [2], Catarina Florindo [3], Alexandra Costa [1], Željko Petrovski [1], Isabel M. Marrucho [3], Rui Pedrosa [2,*] and Luís C. Branco [1,*]

1 LAQV-REQUIMTE, Departamento de Química, Faculdade de Ciências e Tecnologia, Universidade Nova de Lisboa, 2829-516 Caparica, Portugal; amp.costa@campus.fct.unl.pt (A.C.); z.petrovski@fct.unl.pt (Ž.P.)

2 MARE–Marine and Environmental Sciences Centre, ESTM, Instituto Politécnico de Leiria, 2520-641 Peniche, Portugal; celso.alves@ipleiria.pt (C.A.); joana.m.silva@ipleiria.pt (J.S.)

3 Centro de Química Estrutural, Instituto Superior Técnico, Universidade de Lisboa, Avenida Rovisco Pais, 1049-001 Lisboa, Portugal; catarina_florindo@hotmail.com (C.F.); isabel.marrucho@tecnico.ulisboa.pt (I.M.M.)

* Correspondence: miguelmsantos@fct.unl.pt (M.M.S.); rui.pedrosa@ipleiria.pt (R.P.); l.branco@fct.unl.pt (L.C.B.)

Received: 1 June 2020; Accepted: 13 July 2020; Published: 23 July 2020

Abstract: As the development of novel antibiotics has been at a halt for several decades, chemically enhancing existing drugs is a very promising approach to drug development. Herein, we report the preparation of twelve organic salts and ionic liquids (OSILs) from ciprofloxacin and norfloxacin as anions with enhanced antimicrobial activity. Each one of the fluoroquinolones (FQs) was combined with six different organic hydroxide cations in 93–100% yield through a buffer-assisted neutralization methodology. Six of those were isomorphous salts while the remaining six were ionic liquids, with four of them being room temperature ionic liquids. The prepared compounds were not toxic to healthy cell lines and displayed between 47- and 1416-fold more solubility in water at 25 and 37 °C than the original drugs, with the exception of the ones containing the cetylpyridinium cation. In general, the antimicrobial activity against *Klebsiella pneumoniae* was particularly enhanced for the ciprofloxacin-based OSILs, with up to ca. 20-fold decreases of the inhibitory concentrations in relation to the parent drug, while activity against *Staphylococcus aureus* and the commensal *Bacillus subtilis* strain was often reduced. Depending on the cation–drug combination, broad-spectrum or strain-specific antibiotic salts were achieved, potentially leading to the future development of highly bioavailable and safe antimicrobial ionic formulations.

Keywords: active pharmaceutical ingredients as organic salts and ionic liquids (API–OSILs); antibiotics; ciprofloxacin; fluoroquinolones; ionic liquids; norfloxacin; polymorphism

1. Introduction

Fluoroquinolones are a class of antibiotics primarily effective against Gram-negative bacteria. Norfloxacin was the first fluoroquinolone to receive US Food and Drug Administration approval and is currently used to treat urinary, biliary, and respiratory tract infections [1]. One year later, in 1987, ciprofloxacin was also approved for use in the US, and it was the most successful compound in its generation as it showed 4–10 times more efficiency than its predecessor [2]. This drug is very effective in the treatment of a wide range of infections such as urinary tract infection, osteomyelitis (bone infection), respiratory infections and some sexually-transmitted diseases (e.g., gonococcal and chronic bacterial prostatitis) [3].

Fluoroquinolones, like many other active pharmaceutical ingredients (APIs), display spontaneous polymorphic conversion between distinct crystalline forms with a consequent change in pharmaceutical properties [4]. Crystal structures of neutral and zwitterionic forms of norfloxacin and ciprofloxacin have been described by Nikaido and Thanassi, with considerable differences in their bioavailability [5].

In aqueous solutions, a dynamic equilibrium of several protolytic forms (anionic, neutral, zwitterionic and cationic) can be found for both antibiotics due to the presence of distinct functional groups that are prone to converse protonation/deprotonation in physiological media. Such behaviour clearly influences their bioavailability, with neutral ciprofloxacin and norfloxacin displaying 0.086 and 0.37 mg/mL of solubility in water at 25 °C, respectively [6]. The typical conversion to the halide salt renders a considerable increase of two orders of magnitude in water solubility, with the ciprofloxacin hydrochloride salt reaching 38.4 mg/mL at 30 °C [7]. However, it is still considered a poorly water soluble drug.

According to the guidelines of the United States Pharmacopeia [8], solubility improvement techniques of poorly water soluble active pharmaceutical ingredients can be categorised into physical modification, such as chemical modifications of the drug substance, and other techniques. The formation of organic salts of a drug is a mature and explored chemical modification technique.

For more than a decade, active pharmaceutical ingredients as organic salts and ionic liquids (API–OSILs) have risen in academia as an alternative formulation for low bioavailable drugs [9–12]. This third generation of ionic liquids [9] consists on the combination of APIs as cations or anions with organic counterions, thereby inducing distinctive physicochemical properties over the original drugs and reduced toxicity to healthy cells, thus rendering a potentially enhanced pharmaceutical activity in comparison to the API [13–25]. Our works involving the preparation of API–OSILs from, e.g., β-lactam (ampicillin [18–21], penicillin [22] and amoxicillin [22]) and fluoroquinolone [18,23] antibiotics, NSAIDs (ibuprofen [17,18], naproxen [18]), bone antiresorptive agents (zoledronic [24] and alendronic [25] acids), among others, have shown that the combination of an API, either as a cation or as an anion, with suitable biocompatible counter-ions can increase the water solubility of the parent drug and even change its biological effect [10,17,23–28]. Furthermore, the stability and solubility in physiological media of the API can be significantly altered, yielding novel formulations with different pharmacokinetics and potentially different delivery modes and applications [29,30]. In addition, the polymorphic tendency of an API can be considerably reduced or even eliminated (e.g., by attaining the liquid state at room temperature), hence tackling one of the most important issues in the pharmaceutical industry, which is responsible for limitations in a drug's solubility and dosage [17,23,24].

The formation of organic salts based on cationic fluoroquinolones (mostly comprising ciprofloxacin (Cip), but also norfloxacin (Nor)) has been widely explored by the academic community by protonation of the drugs' amine group(s). More precisely, these fluoroquinolones have been combined with dicarboxylic acids (malonic and tartaric [31], adipic, fumaric [32] and maleic [33,34], succinic [35]), citric acid [32], saccharine [36,37] and acesulfame [36], as well as with other drugs (diflunisal and ibuprofen [38], salicylic [39] and barbituric acids [40]). Some works involve the combination of ciprofloxacin or enrofloxacin as cations and/or as anions with amino acids [41] and ionic polymers [42,43]. While most works focus on the study of the chemical and physicochemical properties of the prepared salts, including bioavailability, crystallinity/amorphous profiles and thermodynamic stability, the amorphous–solid dispersions of ciprofloxacin and enrofloxacin showed, in addition, that the antimicrobial activity of the salts is enhanced in comparison with the parent drugs, in particular against Gram-negative bacteria strains.

In addition to these works, our group reported the synthesis and characterization of novel API–OSILs from Cip and Nor as cations in combination with methanesulfonate (Mes), gluconate (Glu) and glycolate (Gly) biocompatible anions [23] (see Figure 1). The bioavailability of the prepared fluoroquinolone-based salts (FQ–OSILs), expressed in terms of permeability (octanol/water and phospholipid/water partition) and solubility in water and biological fluids, was highly tuneable according to the selected anion. Finally, toxicological studies on intestinal epithelial cell line models (Caco-2) revealed a lower inflammatory response for some of the FQ–OSILs in comparison with the parent drugs.

In this work, we set out to explore for the first time the formation of FQ–OSILs based on Cip and Nor as anions in combination with organic cation hydroxides [17,42] based on ammonium, pyridinium and *N*-methylimidazolium moieties (Figure 1).

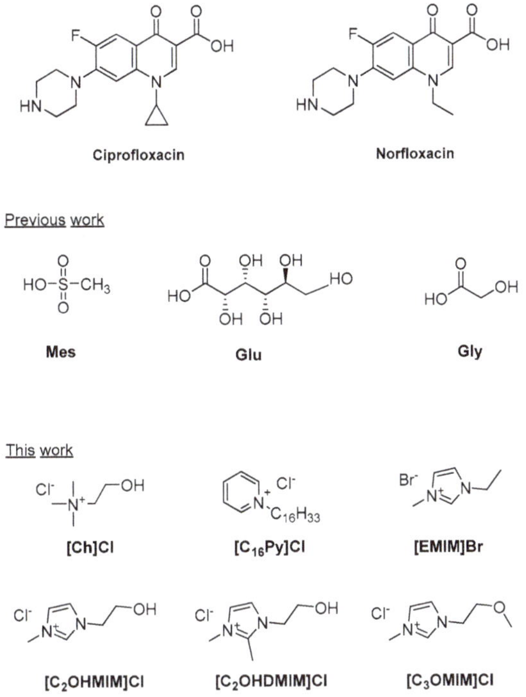

Ciprofloxacin

Norfloxacin

Previous work

Mes

Glu

Gly

This work

[Ch]Cl

$[C_{16}Py]Cl$

[EMIM]Br

$[C_2OHMIM]Cl$

$[C_2OHDMIM]Cl$

$[C_3OMIM]Cl$

Figure 1. Structures of ciprofloxacin and norfloxacin organic acids used as anions in the previous work and organic halide cations used in this work.

The newly synthesized FQ–OSILs were characterised by nuclear magnetic resonance (NMR) (^1H and ^{13}C) and Fourier-transform infrared (FTIR) spectroscopic techniques, elemental analysis, as well as differential scanning calorimetry (DSC). Water solubility at 25 and 37 °C was experimentally determined for the FQ–OSILs, with the exception of the ones containing the $[C_{16}Py]$ cation, for which the critical micelle concentration was measured. Finally, the antimicrobial activity of these anionic FQ–OSILs on Gram-negative (*Klebsiella pneumoniae*) and Gram-positive (*Bacillus subtilis, Staphylococcus aureus*) bacteria was determined.

2. Results and Discussion

2.1. Synthesis

Ciprofloxacin and norfloxacin were combined as anions with choline [Ch], 1-ethyl-3-methylimidazolium [EMIM], 1-hydroxy-ethyl-3-methylimidazolium $[C_2OHMIM]$, 1-(2-hydroxyethyl)-2,3-dimethylimidazolium $[C_2OHDMIM]$, 1-(2-methoxyethyl)-3-methylimidazolium $[C_3OMIM]$ and cetylpyridinium $[C_{16}Py]$ cations (see Scheme 1).

According to our experience [17,24], the most efficient methodology for the combination of APIs as anions with organic cations consists of the dropwise addition of the hydroxide salts of the cations to the neutral API, leading to the sole formation of the desired API–OSILs and water. The cation hydroxides are prepared immediately before their addition from the corresponding halide salts by anion exchange from a hydroxide resin such as Amberlyst IRA-400 (OH). Whenever this methodology is performed with zwitterionic APIs [19,22] such as fluoroquinolones, it is required that the reaction proceeds in ammonia buffer solution so that complete ionisation can occur. The pure API–OSILs were isolated in very high yields (93–100%) after recrystallisation from chloroform/methanol mixtures. Scheme 1 shows the employed synthetic methodology.

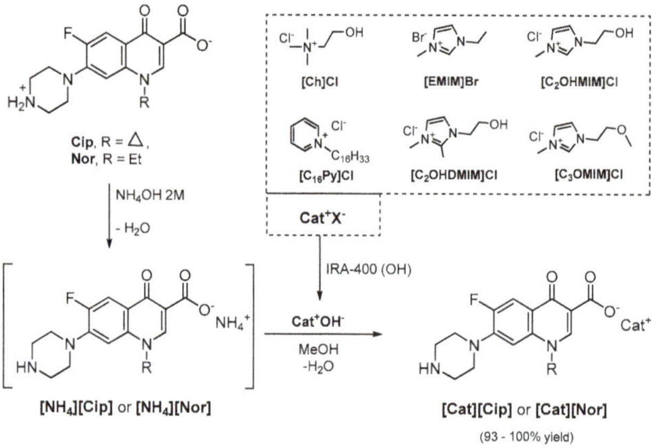

Scheme 1. General synthetic methodology for the synthesis of anionic fluoroquinolone (FQ)-based organic salts and ionic liquids (OSILs).

2.2. Spectroscopic Characterisation

All products were characterised by ^{1}H and ^{13}C NMR and FTIR spectroscopic techniques, as well as elemental analysis.

In all cases, the ^{1}H NMR spectra showed that the cation/anion proportion is strictly 1.0:1.0, in agreement with the intended stoichiometry (see Figures S1–S22). In addition, only one set of signals was observed, meaning that the reactions were complete, and only one product was formed. No comparison with parent ciprofloxacin and norfloxacin is achievable due to a lack of solubility in the same solvent systems as the FQ–OSILs.

The FTIR spectra of the synthesised FQ–OSILs (Figures S23–S35) show only discrete differences in characteristic signals in comparison with those from the parent fluoroquinolones. Typically, one would expect that the deprotonation by the cation hydroxides would occur at the carboxylic acid group forming a carboxylate group, which shows a very different wavenumber in comparison with the initial group. However, this observation was not possible because fluoroquinolones are zwitterions, and consequently, the carboxylic acid group is partially or totally in the form of carboxylate. In our sample of initial ciprofloxacin, the obtained FTIR spectrum showed complete ionisation of the carboxylic acid group, with only the signal from the stretching vibration of O=C-O$^-$ appearing at 1589 cm^{-1} (see Figure 2A).

In this case, the FTIR spectra of the Cip–OSILs showed slight changes in the corresponding vibration band, more precisely between 1581 and 1575 cm^{-1} (e.g., 1575 cm^{-1} for [C$_3$OMIM][Cip] in Figure 2B), thereby suggesting a change in the spatial vicinity of this group consistent with the formation of the salts. In the case of norfloxacin, the obtained FTIR spectrum (Figure 2C) showed incomplete ionisation of the carboxylic acid, and thus two bands were observable at 1727 and 1583 cm^{-1} from the stretching vibration of the carboxylic and carboxylate groups, respectively. The spectra of the Nor–OSILs showed complete disappearance of the signal from the former group, while the latter showed small changes to a range between 1583 and 1579 cm^{-1} (e.g., 1579 cm^{-1} for [C_3OMIM][Nor] in Figure 2D). Furthermore, the 3600–3300 cm^{-1} zone always displays strong signals in the spectra of the final compounds, as opposed to the initial ones, which suggests the establishment of H-bonds between the cations and the anions.

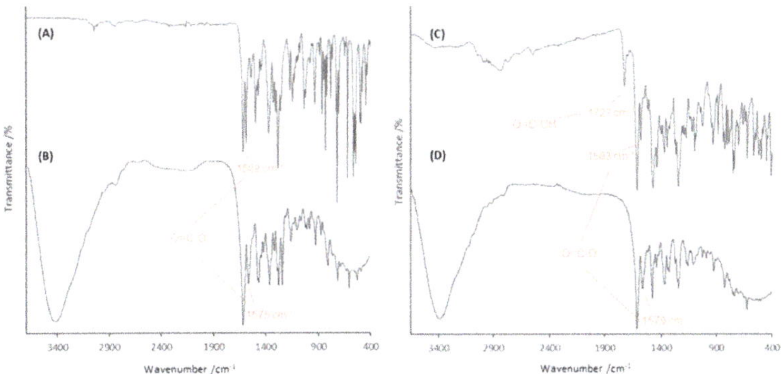

Figure 2. FTIR spectra of (**A**) ciprofloxacin, (**B**) [C_3OMIM][Cip], (**C**) norfloxacin and (**D**) [C_3OMIM][Nor].

2.3. Thermal Analysis

All prepared OSILs containing anionic ciprofloxacin and norfloxacin were studied by differential scanning calorimetry (DSC; see Figures S36–S47). Table 1 contains the obtained data, namely, melting and glass transition temperatures, as well as the physical state at room temperature of the analysed compounds.

Table 1. Physical state at room temperature, melting (T_m) and glass transition (T_g) temperatures of ciprofloxacin, norfloxacin and corresponding OSILs.

Compound	Physical State	T_m/°C	T_g/°C
Ciprofloxacin	White solid	322.0 [44]	-
[Ch][Cip]	Pale yellow solid	111.2	-
[EMIM][Cip]	White solid	92.9	36.7
[C_2OHMIM][Cip]	White solid	111.6	49.5
[C_2OHDMIM][Cip]	White solid	197.5	-
[C_3OMIM][Cip]	White solid	112.4	46.4
[C_{16}Py][Cip]	Orange paste	-	−10.7
Norfloxacin	White solid	217.0 [45]	
[Ch][Nor]	Pale yellow solid	94.5	54.8
[EMIM][Nor]	Yellow solid	119.9	64.6
[C_2OHMIM][Nor]	Yellow paste	-	41.1
[C_2OHDMIM][Nor]	White solid	121.8	65.1
[C_3OMIM][Nor]	White paste	-	44.1
[C_{16}Py][Nor]	Yellow paste	-	5.9

As expected, all compounds display lower melting temperatures than the parent fluoroquinolones. Globally, from the twelve FQ–OSILs, one half are organic salts ([Ch][Cip], [EMIM][Nor], [C_2OHMIM][Cip], [C_2OHDMIM][Cip], [C_2OHDMIM][Nor] and [C_3OMIM][Cip]), while the other half are considered ionic liquids ([Ch][Nor], [EMIM][Cip], [C_2OHMIM][Nor], [C_3OMIM][Nor], [C_{16}Py][Cip] and [C_{16}Py][Nor]) because their melting temperatures are, respectively, higher and lower than 100 °C. All solid compounds display only one melting temperature and are thus considered isomorphic salts. Moreover, the melting temperature range is comprehended between 92.9 and 197.5 °C, respectively, registered for [EMIM][Cip] and [C_2OHDMIM][Cip], and was observed after one isotherm at 80 °C for 15 min, followed by a cooling cycle at 20 °C/min until −90 °C. In the cooling cycles following the melting phenomena, the OSILs do not display crystallization temperatures, meaning that they become supercooled salts, i.e., amorphous, after the first melt. This thermal behaviour is consistent with Type II, as described by Domínguez et al. [46].

From the referred set of ionic liquids, [C_2OHMIM][Nor], [C_3OMIM][Nor], [C_{16}Py][Cip] and [C_{16}Py][Nor] were obtained as room temperature ionic liquids (RTILs), consistent with the observed glass transition temperatures (T_g). Even though several cooling cycles at 10 and 20 °C/min until −90 °C were performed, no endothermic events ascribable to melting phenomena were observed for the RTILs, in agreement with Domínguez's Type I classification [46]. Note that from the four RTILs, three are norfloxacin-based, while only one contains ciprofloxacin. In fact, from the whole set of prepared FQ–OSILs, the melting temperatures of the Nor-based OSILs are invariably lower than the corresponding ones containing Cip. This observation is in line with the T_m of both free fluoroquinolones, with Cip and Nor recording 322 and 217 °C, respectively. Such considerable difference could not be expected by simply observing both chemical structures, as they only differ at the *N*-alkyl substituent (cyclopropyl versus ethyl) at the quinolone moiety. As OSILs, the three-dimensional packing of the fluoroquinolones and additional specific interactions of these drugs with the selected ammonium, methylimidazolium and pyridinium counterions [47] may account for the observed thermal behaviour of the compounds. These observations will be further explored in future works, e.g., by X-ray diffraction techniques, and published accordingly.

2.4. Water Solubility Studies

The solubility of the prepared organic salts based on ciprofloxacin and norfloxacin in water was determined by adding an excess amount of the compound to a fixed mass of water. Briefly, vials were kept under vigorous stirring and controlled temperature, from which samples were collected at different time periods to ensure the equilibrium was reached. After centrifugation, Cip and Nor were quantified through UV–Vis spectroscopy techniques.

All synthesised FQ–OSILs present higher solubility in water than the parent fluoroquinolones, either as free bases or as hydrochloride salts (only for ciprofloxacin). Figure 3 and Table S1 show the data obtained at 25 and 37 °C.

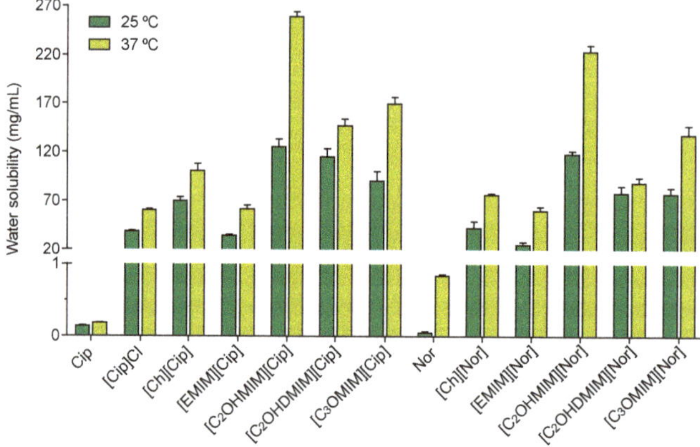

Figure 3. Water solubility at 25 (green) and 37 °C (yellow) for ciprofloxacin, norfloxacin and corresponding OSILs. Values represent mean ± standard error of the mean (SEM) of at least three independent experiments.

On a first observation, it is easily perceived that there is an enhanced increase in water solubility of the ciprofloxacin-based OSILs in comparison with the ones containing norfloxacin. Additionally, there is a direct correlation of the solubility with the temperature.

In both Nor- and Cip-based families of OSILs, the ones with the [EMIM] cation were found to be the least soluble ones, yet increases of 47- and 246-fold at 25 °C and of 72- and 343-fold at 37 °C were, respectively, observed. Conversely, the most soluble OSILs were obtained through the combination of the APIs with [C_2OHMIM], with enhancements between 219- and 897-times at 25 °C, and 266- and 1416-times at 37 °C, respectively, for the Nor- and Cip-containing OSILs.

The observed general tendency is in agreement with the hydrophilic character of the studied cations: [EMIM] < [Ch] < [C_2OHDMIM] ≈ [C_3OMIM] < [C_2OHMIM]. This relative behaviour is in agreement with the previously observed behaviour for ampicillin-based OSILs [20]. On the one hand, the [EMIM] cation does not possess any polar group in its structure, thereby inhibiting its ability to perform strong H-bonding interactions with the anions. On the other hand, the introduction of a polar hydroxyl group at the end of the ethyl group linked to the imidazolium ring in [C_2OHMIM] increased the water solubility by 3.6- and 4.7-times for the corresponding Cip- and Nor-based OSILs. Changing the methylimidazolium moiety by an ammonium group, where the positive charge is more spatially hindered, had a deleterious effect in the solubility of the salts, which decreased by 1.8- and 2.8-times for [Ch][Cip] and [Ch][Nor], respectively, in comparison with the corresponding [C_2OHMIM] ones. With the introduction of a methyl group between the methylimidazolium nitrogen atoms ([C_2OHDMIM]) or at the end of the hydroxyethyl side chain ([C_3OMIM]), intermediate hydrophilic properties are attained, which directly correlates with the solubility in water of the corresponding Cip– and Nor–OSILs.

In general, the anionic fluoroquinolone-based OSILs display a diminished enhancement of the solubility in water in comparison with the cationic fluoroquinolone-based OSILs previously reported [23]. In the latter group, with the exception of [Nor][Gly], the OSILS presented water solubility values between ca. 200 mg/mL up to totally soluble. However, this solubility profile may be attributed to the higher hydrophilic character of the chosen anions in comparison with the set of cations used in this work, and not actually to the anionic or cationic state of the fluoroquinolones.

2.5. Critical Micelle Concentration

The surfactant properties of the ionic liquids formed by the combination of the [C_{16}Py] cation with both fluoroquinolones ([C_{16}Py][Cip] and [C_{16}Py][Nor]) were studied through the calculation of the corresponding critical micelle concentrations (CMCs) via ionic conductivity measurements. This data may be particularly relevant in order to develop novel drug delivery systems [48] for these FQ–OSILs. Table 2 and Figure S48 show the obtained data for both OSILs, as well as for [C_{16}Py]Cl.

Table 2. Critical micelle concentrations (CMCs) of [C_{16}Py]Cl, [C_{16}Py][Cip] and [C_{16}Py][Nor] in water at 25 °C.

Compounds	CMC (mmol/L)
[C_{16}Py]Cl	1.063 (1.067 [49]; 0.96 [50])
[C_{16}Py][Cip]	0.0816
[C_{16}Py][Nor]	0.0539

The CMC value obtained for [C_{16}Py]Cl is in good agreement with the value found in the open literature [49,50]. In comparison with [C_{16}Py]Cl, the calculated CMC values for [C_{16}Py][Cip] and [C_{16}Py][Nor] are two orders of magnitude lower and correspond to a decrease of 92% and 95%, respectively. Despite the fact that smaller CMC values are expected for salts with organic anions in comparison with inorganic ones, such a drastic decrease comes as a surprise. In our previous work on ampicillin–OSILs [20], the combination of the antibiotic with the [C_{16}Py] cation yielded a CMC value of 0.444 mM, which corresponds to a decrease of 58%. When comparing the three systems, we conclude that the level of hydration of Cip and Nor is much lower than that of ampicillin, leading to easier adsorption of these drugs on the surface of [C_{16}Py] micellar structures, thus decreasing the repulsion between the polar groups at much lower concentrations in water.

2.6. Cytotoxicity Studies

The toxicity of the prepared FQ–OSILs and corresponding starting materials, namely, starting fluoroquinolones and halide salts, on 3T3 mouse fibroblasts was determined. Figure 4 plots the relative 3T3 cell viability in the presence of the starting materials and the FQ–OSILs at 10 μM.

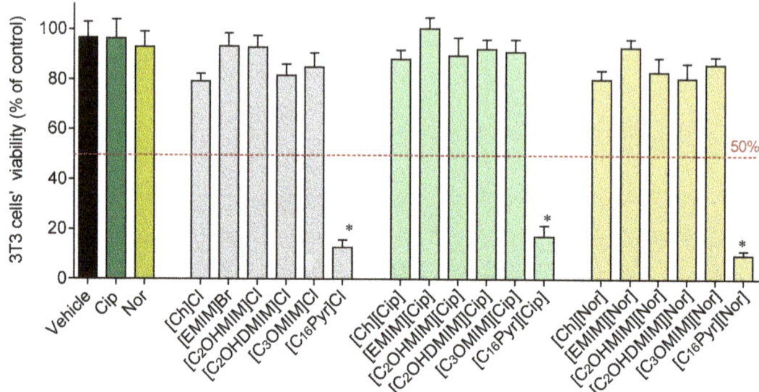

Figure 4. Cytotoxicity of free fluoroquinolones, halide salts and prepared FQ–OSILs on 3T3 cells (10 μM) after treatment for 24 h. Values represent mean ± standard error of the mean (SEM) of at least three independent experiments carried out in triplicate. Symbols represent statistically significant differences (ANOVA, Dunnett's test, p-value < 0.05) when compared to control (*).

From the whole set of compounds, only the ones that contained the [C_{16}Py] cation were found to reduce cell viability by more than 50% at the tested concentration. Therefore, dose–response assays were carried out for these samples. The obtained results are plotted in Figure 5.

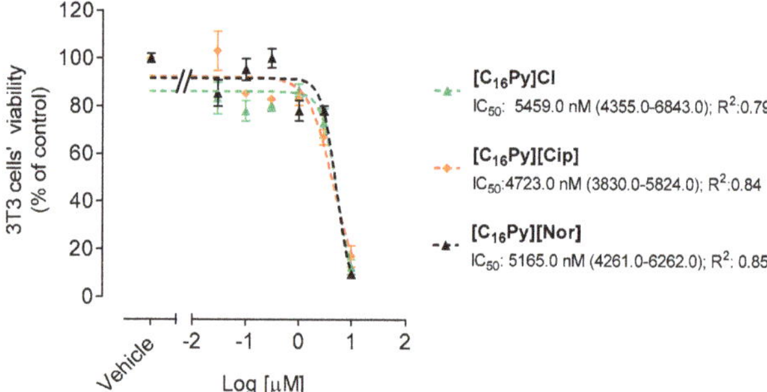

Figure 5. Dose–response curves of [C_{16}Py]-based salts (0.03–10 μM; 24 h) on 3T3 cells for IC_{50} determination. Values represent mean ± standard error of the mean (SEM) of at least three independent experiments carried out in triplicate.

The determined IC_{50} values for [C_{16}Py]Cl, [C_{16}Py][Cip] and [C_{16}Py][Nor] were, respectively, 5.46, 4.72 and 5.17 μM. These values were expected due to the surfactant properties of the cation and are in agreement with previous works [17,51].

2.7. Antimicrobial Activity Studies

The antimicrobial activity (IC_{50}) of the prepared FQ–OSILs and the corresponding starting materials was determined against Gram-negative (*Klebsiella pneumoniae*) and Gram-positive (*Staphylococcus aureus, Bacillus subtilis*) bacteria.

Figure 6 plots the relative growth of (A) *Klebsiella pneumoniae*, (B) *Staphylococcus aureus* and (C) *Bacillus subtilis* in the presence of the starting materials and the FQ–OSILs at 10 μM for 6 h.

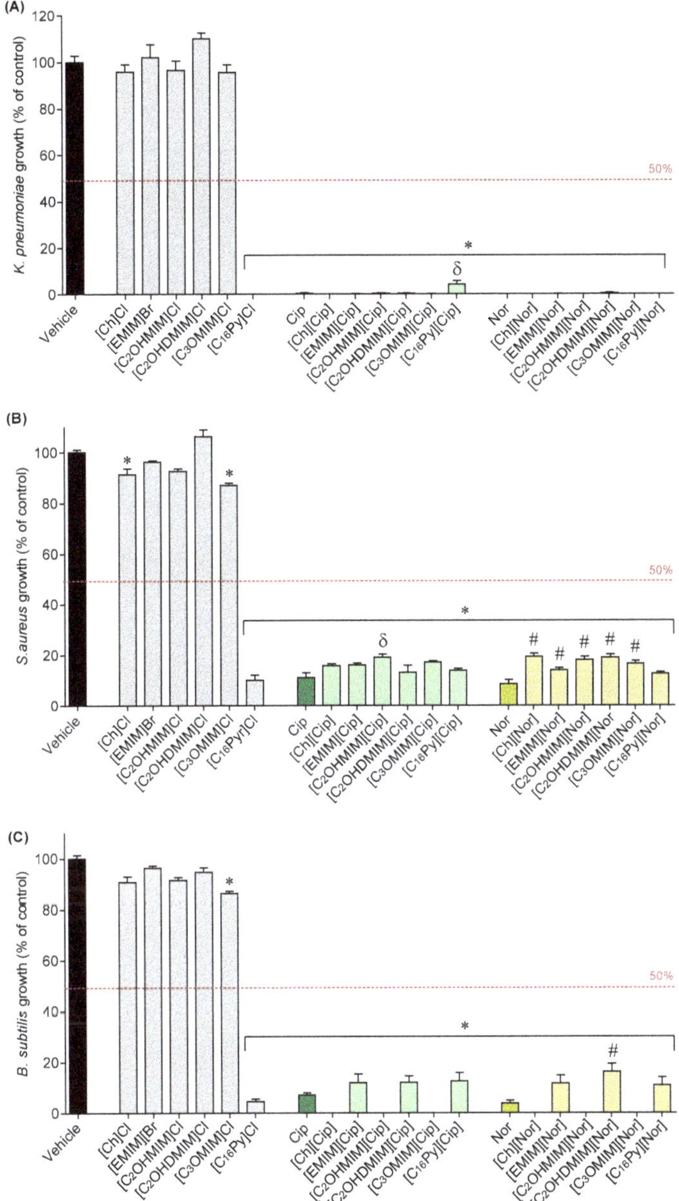

Figure 6. Growth inhibition (**A**) *Klebsiella pneumonia*, (**B**) *Staphylococcus aureus* and (**C**) *Bacillus subtilis* after exposure to the free fluoroquinolones, starting halide salts and prepared FQ–OSILs at 10 μM for 6 h. Values represent mean ± standard error of the mean (SEM) of at least three independent experiments carried out in triplicate. Symbols represent statistically significant differences (ANOVA, Dunnett's test, p-value < 0.05) between the FQ–OSILs and control (*), Cip (δ) and Nor (#).

Both free fluoroquinolones and all prepared FQ–OSILs inhibited more than 50% growth of all bacteria strains at the tested concentration (10 µM). From the set of starting halide cations, [C_{16}Py]Cl was the only one that exerted an effective antimicrobial effect.

Upon performing dose–response assays for the active compounds (see Figure S49), the corresponding IC_{50} values were determined. Tables 3 and 4 present the gathered data for Cip- and Nor-based OSILs (and the corresponding free fluoroquinolones), respectively. The data for the starting cation halides are presented in Table S2.

Table 3. Antimicrobial activity (IC_{50}) values (nM) and relative decrease of inhibitory concentrations ($RDIC_{50}$) for ciprofloxacin and corresponding OSILs (0.001–10 µM) against the tested microorganisms (in **bold** are the most relevant IC_{50} and $RDIC_{50}$ values). The values in parentheses represent the confidence intervals for 95%.

Compounds	*K. pneumoniae*	*RDIC*$_{50}$	*S. aureus*	*RDIC*$_{50}$	*B. subtilis*	*RDIC*$_{50}$
Cip	196.50 (119.6–322.8)		29.16 (20.28–41.92)		3.84 (0.25–58.16)	
[Ch][Cip]	**51.12** (38.44–67.99)	**3.8**	181.4 (122.9–267.9)	0.2	**20.60** (18.86–22.49)	**0.2**
[EMIM][Cip]	**64.80** (56.61–74.17)	**3.0**	**24.24** (15.79–37.19)	**1.2**	**16.17** (12.10–21.60)	**0.2**
[C_2OHMIM][Cip]	**36.42** (32.63–40.66)	**5.4**	61.95 (34.10–112.50)	0.5	**15.75** (14.47–17.16)	**0.2**
[C_2OHDMIM][Cip]	**50.99** (44.04–59.04)	**3.9**	**21.20** (18.03–24.93)	**1.4**	**12.35** (8.31–18.35)	**0.3**
[C_3OMIM][Cip]	**47.61** (39.64–57.19)	**4.1**	82.05 (48.99–137.40)	0.4	**21.66** (17.12–27.40)	**0.2**
[C_{16}Py][Cip]	**9.88** (4.21–23.17)	**19.9**	1430 (1140.0–1793.0)	0.0	**6.07** (5.21–7.06)	**0.6**

Ciprofloxacin was found to be particularly active against *B. subtilis* (3.84 nM), followed by *S. aureus* (29.16 nM) and *K. pneumoniae* (196.5 nM), as expected.

The combination of this antibiotic with the selected cations has rendered very interesting results against these bacteria strains, mostly *K. pneumoniae*, as similarly observed in previous works by Tajber and co-workers [42,43]. More specifically, all Cip–OSILs showed increased activity against this Gram-negative bacteria strain, with IC_{50} values comprehended between 9.88 ([C_{16}Py][Cip]) and 64.8 nM ([EMIM][Cip]), which correspond to a remarkable relative decrease of inhibitory concentrations ($RDIC_{50}$) of 19.9 and 3.0, respectively. While the antimicrobial activity of the former OSIL may come from an additive mechanism—because the [C_{16}Py] cation itself shows an IC_{50} value of 2.55 mM (see Table S2)—the potency of the remaining OSILs appears to have a synergistic behaviour since none of the cations display activity against this strain (or any of the others). In spite of [C_{16}Py][Cip] showing toxicity against fibroblasts (see above), its IC_{50} value against *K. pneumoniae* is ca. 500-times lower (4.72 µM vs. 9.88 nM), making this Cip–OSIL particularly interesting against this strain. Moreover, [C_{16}Py][Cip] showed almost no activity against *S. aureus* and also a 40% decrease against the commensal *B. subtilis* in comparison with ciprofloxacin. Hence, these results clearly suggest that [C_{16}Py][Cip] has a very specific activity against *K. pneumoniae* over the remaining tested strains.

Of note is that *B. subtilis* is part of the human gut microbial ecosystem, whose perturbation may lead to the proliferation of resistant bacterial pathogens [52]. In fact, all Cip–OSILs had particularly reduced activities (20–30%) against *B. subtilis* in comparison with ciprofloxacin.

Regarding the pathogenic *S. aureus*, another Gram-positive bacteria strain, only [EMIM][Cip] and [C_2OHDMIM][Cip] showed a slight increase in antimicrobial activity, with $RDIC_{50}$ values between 1.2 and 1.4.

In its turn, free norfloxacin shows that, similarly to ciprofloxacin, it is also more active against Gram-positive bacteria (*S. aureus* and *B. subtilis*) than Gram-negative (*K. pneumoniae*) strains, but to a smaller extent than its fluoroquinolone sibling. Table 4 shows this data, as well as the one observed for the Nor-based OSILs.

Table 4. IC_{50} values (nM) and relative decrease of inhibitory concentrations for 50% activity ($RDIC_{50}$) for norfloxacin and corresponding OSILs (0.001–10 μM) against the tested microorganisms (in **bold** are the most relevant IC_{50} and $RDIC_{50}$ values). The values in parentheses represent the confidence intervals for 95%.

Compounds	*K. pneumoniae*	$RDIC_{50}$	*S. aureus*	$RDIC_{50}$	*B. subtilis*	$RDIC_{50}$
Nor	255.2 (188.6–345.3)		123.4 (71.29–213.5)		117.2 (97.70–140.60)	
[Ch][Nor]	199.4 (179.9–221.0)	1.3	127.1 (85.17–189.60)	1.0	95.49 (80.53–113.2)	1.2
[EMIM][Nor]	207.5 (185.1–232.5)	1.2	211.2 (152.6–292.3)	0.6	24.48 (19.36–30.94)	4.8
[C_2OHMIM][Nor]	201.0 (182.6–221.1)	1.3	91.70 (54.08–155.5)	1.3	88.61 (81.35–96.51)	1.3
[C_2OHDMIM][Nor]	**178.2** (156.1–203.3)	**1.4**	**79.55** (69.24–91.40)	**1.6**	**148.8** (126.1–175.7)	**0.8**
[C_3OMIM][Nor]	224.5 (189.5–266.0)	1.1	106.7 (70.13–162.5)	1.1	68.12 (56.55–82.06)	1.7
[C_{16}Py][Nor]	202.9 (173.9–236.7)	1.3	**10.84** (2.89–40.67)	**11.4**	87.29 (76.04–100.2)	1.3

The most active Nor-based OSIL is, once again, the one containing the [C_{16}Py] cation, which showed a relative increase of 11.4-times (IC_{50} = 10.84 nM) against *S. aureus* in comparison with the free norfloxacin. This result is quite peculiar, as the analogous salt based on ciprofloxacin had similar activity against the Gram-negative *K. pneumoniae* microbial strain. Against the latter, [C_{16}Py][Nor] has shown only a 30% increase in activity, with an IC_{50} value of 202.9 nM, which is much higher than the one recorded on *S. aureus*. However, it negatively affects commensal *B. subtilis* to a higher extent than norfloxacin.

An alternative promising broad-spectrum formulation may be [C_2OHDMIM][Nor], which showed a 60% increased activity against *S. aureus* and 40% against *K. pneumoniae*. Desirably, it also lost 20% activity against commensal *B. subtilis*.

3. Conclusions

The ciprofloxacin- and norfloxacin-based OSILs presented in this paper clearly demonstrate that, by a simple and effective chemical transformation of standard fluoroquinolone drugs, it is possible to enhance their bioavailability and mitigate their polymorphic profiles. These advantages, allied to the possibility to modulate their antimicrobial spectrum while maintaining low toxicity towards healthy cells, make these FQ–OSILs a very promising tool to prepare novel effective formulations for these drugs.

In general, the prepared ciprofloxacin-based OSILs had much more encouraging antimicrobial data in comparison with the norfloxacin ones, despite both sets having positive results. According to the data regarding the three tested bacteria strains (*K. pneumoniae*, *S. aureus* and *B. subtilis*), we postulate that it is possible to develop highly bioavailable tailored formulations of both fluoroquinolones for applications against one or more pathogenic microbes with low toxicity against healthy cells and commensal bacteria, depending on the chosen combination with organic cations. On the one hand, the [C_{16}Py]-based OSILs containing ciprofloxacin and norfloxacin were found to be particularly selective against different bacteria strains, respectively, *K. pneumoniae* and *S. aureus*, both at subtoxic concentrations. Such distinct behaviour must surge from distinct interactions between the cation, the drugs and the cell external membrane/wall of the bacteria, and will be studied in the future. In addition, all formulations based on the [C_{16}Py] cation assembled into micelles, with possible applications in novel drug delivery systems of the antibiotics.

Alternatively, highly water-soluble formulations with a selective spectrum of activity could consist of [Ch][Cip], which was shown to be ca. 3.5-times more active to this strain than to *S. aureus*. However, its IC_{50} values against *K. pneumoniae* are some of the lowest recorded (51.12 nM). Hence, promising alternative formulations can consist of [C_2OHMIM]- and [C_3OMIM][Cip], which demonstrated selectivity of ca. 1.7-times towards the same strain, with IC_{50} values of 36.43 and 47.61 nM, respectively. Moreover, both latter compounds are more water-soluble than [Ch][Cip] and probably more bioavailable.

On the other hand, if a broad-spectrum antibiotic ionic formulation is sought, its combination with [EMIM] or [C_2OHDMIM] is promising as they are highly active against both Gram-negative and Gram-positive bacteria strains, in particular *K. pneumoniae* and *S. aureus*, respectively.

Supplementary Materials: The following are available online at http://www.mdpi.com/1999-4923/12/8/694/s1: Materials, experimental procedures, compound characterization data. Figures S1–S22: ^{1}H and ^{13}C NMR spectra of FQ–OSILs. Figures S23–S35: FTIR spectra of FQ–OSILs. Figures S36–S47: DSC thermograms of FQ–OSILs. Figure S48: Plot of ionic conductivity in function of the concentration of [C_{16}Py][Cip] and [C_{16}Py][Nor]. Figure S49: Dose–response curves of ciprofloxacin, norfloxacin and corresponding FQ–OSILs on *K. pneumoniae S. aureus* and *B. subtilis* for IC_{50} determination. Table S1: Solubility in water of ciprofloxacin, ciprofloxacin hydrochloride, norfloxacin and organic salts based in fluoroquinolones (mean and standard deviation) at 25 and 37 °C. Table S2: IC_{50} values (nM) for the starting halide cations (0.001–10 µM) against the tested microorganisms.

Author Contributions: Conceptualisation, L.C.B. and Ž.P.; methodology and investigation, M.M.S., C.A., J.S., C.F., and A.C.; validation, M.M.S., I.M.M., R.P., and L.C.B; writing—original draft preparation, M.M.S., C.A., and C.F.; writing—review and editing, M.M.S., C.A., I.M.M., and L.C.B.; supervision, I.M.M., R.P., and L.C.B.; funding acquisition, I.M.M., R.P., and L.C.B. All authors have read and agreed to the published version of the manuscript.

Funding: This work was supported by the Associate Laboratory for Green Chemistry LAQV and Centro de Química Estrutural, which are financed by national funds from FCT/MCTES (UIDB/50006/2020 and UID/QUI/00100/2013, respectively). The authors also thank Fundação para a Ciência e Tecnologia for the projects PTDC/QUI-QOR/32406/2017, PEst-C/LA0006/2013, and RECI/BBBBQB/0230/2012, as well as one contract under Investigador FCT (L. C. Branco).

Acknowledgments: The authors acknowledge Professor Madalena Dionísio for the use of the DSC equipment, Bayer for the donation of ciprofloxacin and Solchemar for the cation halides.

Conflicts of Interest: The authors declare no conflict of interest.

Abbreviations

API	active pharmaceutical ingredient
API–OSILs	active pharmaceutical ingredient organic salts and ionic liquids
C_{16}Py	cetylpyridinium
C_2OHDMIM	1-(2-hydroxyethyl)-2,3-dimethylimidazolium
C_2OHMIM	1-(2-hydroxyethyl)-3-methylimidazolium
C_3OMIM	1-(2-methoxyethyl)-3-methylimidazolium
Ch	choline
Cip	ciprofloxacin
CMC	critical micelle concentration
EMIM	1-ethyl-3-methylimidazolium
FQ	fluoroquinolone
FQ-OSILs	fluoroquinolone-based organic salts and ionic liquids
FTIR	Fourier-transform infrared spectroscopy
IC_{50}	half maximal inhibitory concentration
IL	ionic liquid
NMR	nuclear magnetic resonance
Nor	norfloxacin
OSIL	organic salt and ionic liquid
$RDIC_{50}$	relative decrease in inhibitory concentration for 50% bacterial growth
RTILsSEM	room temperature ionic liquidsstandard error mean
T_g	glass transition temperature
T_m	melting temperature

References

1. Holmes, B.; Brogden, R.N.; Richards, D.M. Norfloxacin. A review of its antibacterial activity, pharmacokinetic properties and therapeutic use. *Drugs* **1985**, *30*, 482–513. [CrossRef] [PubMed]
2. Sharma, P.C.; Jain, A.; Jain, S.; Pahwa, R.; Yar, M.S. Ciprofloxacin: Review on developments in synthetic, analytical, and medicinal aspects. *J. Enz. Inhib. Med. Chem.* **2010**, *25*, 577–589. [CrossRef] [PubMed]
3. Emmerson, A.M.; Jones, A.M. The quinolones: Decades of development and use. *J. Antimicrob. Chemother.* **2003**, *51*, 13–20. [CrossRef] [PubMed]
4. Variankaval, N.; Cote, A.S.; Doherty, M.F. From form to function: Crystallization of active pharmaceutical ingredients. *AIChE J.* **2008**, *54*, 1682–1688. [CrossRef]
5. Nikaido, H.; Thanassi, D.G. Penetration of lipophilic agents with multiple protonation sites into bacterial cells: Tetracyclines and fluoroquinolones as examples. *Antimicrob. Agents Chemother.* **1993**, *37*, 1393–1399. [CrossRef]
6. Yu, X.; Zipp, G.; Davidson, G.W.R., III. The effect of temperature and pH on the solubility of quinolone compounds: Estimation of heat of fusion. *Pharm. Res.* **1994**, *11*, 522–527. [CrossRef]
7. Varanda, F.; Pratas de Melo, M.J.; Caço, A.I.; Dohrn, R.; Makrydaki, F.A.; Voutsas, E.; Tassios, D.; Marrucho, I.M. Solubility of antibiotics in different solvents. 1. Hydrochloride forms of tetracycline, moxifloxacin, and ciprofloxacin. *Ind. Eng. Chem. Res.* **2006**, *45*, 6368–6374. [CrossRef]
8. Savjani, K.T.; Gajjar, A.K.; Savjani, J.K. Drug Solubility: Importance and Enhancement Techniques. *Pharmaceutics* **2012**, 195727. [CrossRef]
9. Hough, W.L.; Smiglak, M.; Rodriguez, H.; Swatloski, R.P.; Spear, S.K.; Daly, D.T.; Pernak, J.; Grisel, J.E.; Carliss, R.D.; Soutullo, M.D.; et al. The third evolution of ionic liquids: Active pharmaceutical ingredients. *New J. Chem.* **2007**, *31*, 1429–1436. [CrossRef]
10. Egorova, K.S.; Gordeev, E.G.; Ananikov, V.P. Biological Activity of Ionic Liquids and Their Application in Pharmaceutics and Medicine. *Chem. Rev.* **2017**, *117*, 10–7132. [CrossRef]
11. Ferraz, R.; Branco, L.C.; Prudencio, C.; Noronha, J.P.; Petrovski, Z. Ionic Liquids as Active Pharmaceutical Ingredients. *ChemMedChem* **2011**, *6*, 975–985. [CrossRef] [PubMed]
12. Marrucho, I.M.; Branco, L.C.; Rebelo, L.P.N. Ionic Liquids in Pharmaceutical Applications. *Annu. Rev. Chem. Biomol. Eng.* **2014**, *5*, 527–546. [CrossRef]
13. Smiglak, M.; Pringle, J.M.; Lu, X.; Han, L.; Zhang, S.; Gao, H.; MacFarlane, D.R.; Rogers, R.D. Ionic liquids for energy, materials, and medicine. *Chem. Commun.* **2014**, *50*, 9228–9250. [CrossRef] [PubMed]
14. Shamshina, J.L.; Kelley, S.P.; Gurau, G.; Rogers, R.D. Chemistry: Develop ionic liquid drugs. *Nature* **2015**, *528*, 188–189. [CrossRef] [PubMed]
15. Cojocaru, O.A.; Bica, K.; Gurau, G.; Narita, A.; McCrary, P.D.; Shamshina, J.L.; Barber, P.S.; Rogers, R.D. Prodrug ionic liquids: Functionalizing neutral active pharmaceutical ingredients to take advantage of the ionic liquid form. *MedChemComm* **2013**, *4*, 559–563. [CrossRef]
16. Cherukuvada, S.; Nangia, A. Polymorphism in an API ionic liquid: Ethambutol dibenzoate trimorphs. *CrystEngComm* **2012**, *14*, 7840–7843. [CrossRef]
17. Santos, M.M.; Raposo, L.R.; Carrera, G.V.S.M.; Costa, A.; Dionísio, M.; Baptista, P.V.; Fernandes, A.R.; Branco, L.C. Ionic Liquids and Salts from Ibuprofen as Promising Innovative Formulations of an Old Drug. *ChemMedChem* **2019**, *14*, 907–910. [CrossRef]
18. Carrera, G.V.S.M.; Santos, M.M.; Costa, A.; Rebelo, L.P.N.; Marrucho, I.M.; Ponte, M.N.; Branco, L.C. Highly water soluble room temperature superionic liquids of APIs. *New J. Chem.* **2017**, *41*, 6986–6990. [CrossRef]
19. Ferraz, R.; Branco, L.C.; Marrucho, I.; Araújo, J.; da Ponte, M.N.; Prudêncio, C.; Noronha, J.P.; Petrovski, Z. Development of Novel Ionic Liquids-APIs based on Ampicillin derivatives. *Med. Chem. Commun.* **2012**, *3*, 494–497. [CrossRef]
20. Florindo, C.; Araujo, J.M.M.; Alves, F.; Matos, C.; Ferraz, R.; Prudencio, C.; Noronha, J.P.; Petrovski, Z.; Branco, L.; Rebelo, L.P.N.; et al. Evaluation of solubility and partition properties of ampicillin-based ionic liquids. *Int. J. Pharm.* **2013**, *456*, 553–559. [CrossRef]
21. Ferraz, R.; Teixeira, V.; Rodrigues, D.; Fernandes, R.; Prudencio, C.; Noronha, J.P.; Petrovski, Z.; Branco, L.C. Antibacterial activity of Ionic Liquids based on ampicillin against resistant bacteria. *RSC Adv.* **2014**, *4*, 4301–4307. [CrossRef]

22. Ferraz, R.; Silva, D.; Dias, A.R.; Dias, V.; Santos, M.M.; Pinheiro, L.; Prudêncio, C.; Noronha, J.P.; Petrovski, Z.; Branco, L.C. Synthesis and Antibacterial Activity of Ionic Liquids and Organic Salts based on Penicillin G and Amoxicillin hydrolysate derivatives against Resistant Bacteria. *Pharmaceutics* **2020**, *12*, 221. [CrossRef] [PubMed]
23. Florindo, C.; Costa, A.; Matos, C.; Nunes, S.L.; Matias, A.N.; Duarte, C.M.M.; Rebelo, L.P.N.; Branco, L.C.; Marrucho, I.M. Novel organic salts based on fluoroquinolone drugs: Synthesis, bioavailability and toxicological profiles. *Int. J. Pharm.* **2014**, *469*, 179–189. [CrossRef] [PubMed]
24. Teixeira, S.; Santos, M.M.; Ferraz, R.; Prudêncio, C.; Fernandes, M.H.; Costa-Rodrigues, J.; Branco, L.C. A Novel Approach for Bisphosphonates: Ionic Liquids and Organic Salts from Zoledronic Acid. *ChemMedChem* **2019**, *14*, 1767–1770. [CrossRef]
25. Teixeira, S.; Santos, M.M.; Ferraz, R.; Prudêncio, C.; Fernandes, M.H.; Costa-Rodrigues, J.; Branco, L.C. Alendronic acid as ionic liquid: New perspective on osteosarcoma. *Pharmaceutics* **2020**, *12*, 293. [CrossRef]
26. Araújo, J.M.M.; Florindo, C.; Pereiro, A.B.; Vieira, N.S.M.; Matias, A.A.; Duarte, C.M.M.; Rebelo, L.P.N.; Marrucho, I.M. Cholinium-based ionic liquids with pharmaceutically active anions. *RSC Adv.* **2014**, *4*, 28126–28132.
27. McCrary, P.D.; Beasley, P.A.; Gurau, G.; Narita, A.; Barber, P.S.; Cojocaru, O.A.; Rogers, R.D. Drug specific, tuning of an ionic liquid's hydrophilic–lipophilic balance to improve water solubility of poorly soluble active pharmaceutical ingredients. *New J. Chem.* **2013**, *37*, 2196–2202. [CrossRef]
28. Ferraz, R.; Costa-Rodrigues, J.; Fernandes, M.H.; Santos, M.M.; Marrucho, I.M.; Rebelo, L.P.; Prudêncio, C.; Noronha, J.P.; Petrovski, Ž.; Branco, L.C. Antitumor Activity of Ionic Liquids Based on Ampicillin. *ChemMedChem* **2015**, *10*, 1480–1483. [CrossRef]
29. Wu, H.; Deng, Z.; Zhou, B.; Qi, M.; Hong, M.; Ren, G. Improved transdermal permeability of ibuprofen by ionic liquid technology: Correlation between counterion structure and the physicochemical and biological properties. *J. Mol. Liq.* **2019**, *283*, 399–409. [CrossRef]
30. Sidat, Z.; Marimuthu, T.; Kumar, P.; du Toit, L.C.; Kondiah, P.P.D.; Choonara, Y.E.; Pillay, V. Ionic Liquids as Potential and Synergistic Permeation Enhancers for Transdermal Drug Delivery. *Pharmaceutics* **2019**, *11*, 96. [CrossRef]
31. Reddy, J.S.; Ganesh, S.V.; Nagalapalli, R.; Dandela, R.; Solomon, K.A.; Kumar, K.A.; Goud, N.R.; Nangia, A. Fluoroquinolone salts with carboxylic acids. *J. Pharm. Sci.* **2011**, *100*, 3160–3176. [CrossRef] [PubMed]
32. Surov, A.O.; Voronin, A.P.; Drozd, K.V.; Churakov, A.V.; Roussel, P.; Perlovich, G.L. Diversity of crystal structures and physicochemical properties of ciprofloxacin and norfloxacin salts with fumaric acid. *CrystEngComm* **2018**, *20*, 755–767. [CrossRef]
33. Surov, A.O.; Manin, A.N.; Voronin, A.P.; Drozd, K.V.; Simagina, A.A.; Churakov, A.V.; Perlovich, G.L. Pharmaceutical salts of ciprofloxacin with dicarboxylic acids. *Eur. J. Pharm. Sci.* **2015**, *77*, 112–121. [CrossRef]
34. Surov, A.O.; Churakov, A.V.; Perlovich, G.L. Three Polymorphic Forms of Ciprofloxacin Maleate: Formation Pathways, Crystal Structures, Calculations, and Thermodynamic Stability Aspects. *Cryst. Growth Des.* **2016**, *16*, 6556–6567. [CrossRef]
35. Paluch, K.J.; McCabe, T.; Müller-Bunz, H.; Corrigan, O.I.; Healy, A.M.; Tajber, L. Formation and physicochemical properties of crystalline and amorphous salts with different stoichiometries formed between ciprofloxacin and succinic acid. *Mol. Pharm.* **2013**, *10*, 3640–3654. [CrossRef] [PubMed]
36. Garro Linck, Y.; Chattah, A.K.; Graf, R.; Romanuk, C.B.; Olivera, M.E.; Manzo, R.H.; Monti, G.A.; Spiess, H.W. Multinuclear solid state NMR investigation of two polymorphic forms of ciprofloxacin-saccharinate. *Phys. Chem. Chem. Phys.* **2011**, *13*, 6590–6596. [CrossRef]
37. Basavoju, S.; Bostrm, D.; Velaga, S.P. Pharmaceutical salts of fluoroquinolone antibacterial drugs with acesulfame sweetener. *Mol. Cryst. Liq. Cryst.* **2012**, *562*, 254–264. [CrossRef]
38. Bag, P.P.; Ghosh, S.; Khan, H.; Devarapalli, R.; Reddy, C.M. Drug–drug salt forms of ciprofloxacin with diflunisal and indoprofen. *CrystEngComm* **2014**, *16*, 7393–7396. [CrossRef]
39. Surov, A.O.; Vasilev, N.A.; Churakov, A.V.; Stroh, J.; Emmerling, F.; Perlovich, G.L. Solid Forms of Ciprofloxacin Salicylate: Polymorphism, Formation Pathways, and Thermodynamic Stability. *Cryst. Growth Des.* **2019**, *19*, 2979–2990. [CrossRef]
40. Golovnev, N.N.; Molokeev, M.S.; Lesnikov, M.K.; Atuchin, V.V. Two salts and the salt cocrystal of ciprofloxacin with thiobarbituric and barbituric acids: The structure and properties. *J. Phys. Org. Chem.* **2018**, *31*, 1–11. [CrossRef]

41. Mesallati, H.; Conroy, D.; Hudson, S.; Tajber, L. Preparation and Characterization of Amorphous Ciprofloxacin-Amino Acid Salts. *Eur. J. Pharm. Biopharm.* **2017**, *121*, 73–89. [CrossRef] [PubMed]
42. Mesallati, H.; Umerska, A.; Paluch, K.J.; Tajber, L. Amorphous Polymeric Drug Salts as Ionic Solid Dispersion Forms of Ciprofloxacin. *Mol. Pharm.* **2017**, *14*, 2209–2223. [CrossRef] [PubMed]
43. Mesallati, H.; Umerska, A.; Tajber, L. Fluoroquinolone Amorphous Polymeric Salts and Dispersions for Veterinary Uses. *Pharmaceutics* **2019**, *11*, 268. [CrossRef] [PubMed]
44. Al-Omar, M.A. Ciprofloxacin: Physical Profile. *Profiles Drug Subst. Excip. Related Methodol.* **2005**, *31*, 163–178.
45. Mendes, C.; Wiemes, B.P.; Buttchevitz, A.; Christ, A.P.; Ribas, K.G.; Adams, A.I.H.; Silva, M.A.S.; Oliveira, P.R. Investigation of β-cyclodextrin–norfloxacin inclusion complexes. Part 1. Preparation, physicochemical and microbiological characterization. *Expert Rev. Anti-Infect. Ther.* **2015**, *13*, 119–129. [CrossRef]
46. Gómez, E.; Calvarn, N.; Domínguez, A. Thermal Behavior of Pure Ionic Liquids. In *Ionic Liquids: Current State of the Art*; Handy, S., Ed.; Intech: Rijeka, Croatia, 2015; Volume 8, pp. 199–228.
47. Mokadem, K.; Korichi, M.; Tumba, K. An enhanced group-interaction contribution method for the prediction of glass transition temperature of ionic liquids. *Fluid Phase Equilibria* **2016**, *425*, 259–268. [CrossRef]
48. Li, C.; Wang, J.; Wang, Y.; Gao, H.; Wei, G.; Huang, Y.; Yu, H.; Gan, Y.; Wang, Y.; Mei, L.; et al. Recent progress in drug delivery. *Acta Pharm. Sinica B* **2019**, *9*, 1145–1162. [CrossRef]
49. Wang, K.; Karlsson, G.; Almgren, M.; Asakawa, T. Aggregation behavior of cationic fluorosurfactants in water and salt solutions. A cryoTEM survey. *J. Phys. Chem. B* **1999**, *103*, 9237–9246. [CrossRef]
50. Chatterjee, A.; Moulik, S.P.; Sanyal, S.K.; Mishra, B.K.; Puri, P.M. Thermodynamics of Micelle Formation of Ionic Surfactants: A Critical Assessment for Sodium Dodecyl Sulfate, Cetyl Pyridinium Chloride and Dioctyl Sulfosuccinate (Na Salt) by Microcalorimetric, Conductometric, and Tensiometric Measurements. *J. Phys. Chem. B* **2001**, *105*, 12823–12831. [CrossRef]
51. Kim, Y.J.; Rossa, C., Jr.; Kirkwood, K.L. Prostaglandin Production by Human Gingival Fibroblasts Inhibited by Triclosan in the Presence of Cetylpyridinium Chloride. *J. Periodon.* **2005**, *76*, 1735–1742. [CrossRef]
52. Wong, W.F.; Santiago, M. Microbial approaches for targeting antibiotic-resistant bacteria. *Microb. Biotechnol. 2017 10*, 1047–1053. [CrossRef]

Article

Ionic Liquid Forms of the Antimalarial Lumefantrine in Combination with LFCS Type IIIB Lipid-Based Formulations Preferentially Increase Lipid Solubility, In Vitro Solubilization Behavior and In Vivo Exposure

Erin Tay [1,2], Tri-Hung Nguyen [2], Leigh Ford [1,3], Hywel D. Williams [3], Hassan Benameur [4], Peter J. Scammells [1,*] and Christopher J. H. Porter [2,5,*]

1 Medicinal Chemistry, Monash Institute of Pharmaceutical Sciences, Monash University, 381 Royal Parade, Parkville, Victoria 3052, Australia; Erin.tay@monash.edu (E.T.); leigh.ford@lonza.com (L.F.)
2 Drug Delivery, Disposition and Dynamics, Monash Institute of Pharmaceutical Sciences, Monash University, 381 Royal Parade, Parkville, Victoria 3052, Australia; tri-hung.nguyen@monash.edu
3 Oral Drug Delivery Innovation, Lonza Pharma Biotech & Nutrition, Monash Institute of Pharmaceutical Sciences, Monash University, 381 Royal Parade, Parkville, Victoria 3052, Australia; Hywel.williams@lonza.com
4 Oral Drug Delivery Innovation, Lonza Pharma Biotech & Nutrition, 67412 Strasbourg, France; hassan.benameur@lonza.com
5 ARC Centre of Excellence in Convergent Bio-Nano Science and Technology, Monash University, 381 Royal Parade, Parkville, Victoria 3052, Australia
* Correspondence: peter.scammells@monash.edu (P.J.S.); chris.porter@monash.edu (C.J.H.P.); Tel.: +61-(0)-3-9903-9542 (P.J.S.); +61-(0)-3-9903-9549 (C.J.H.P.)

Received: 28 November 2019; Accepted: 18 December 2019; Published: 22 December 2019

Abstract: Lipid based formulations (LBFs) are commonly employed to enhance the absorption of highly lipophilic, poorly water-soluble drugs. However, the utility of LBFs can be limited by low drug solubility in the formulation. Isolation of ionizable drugs as low melting, lipophilic salts or ionic liquids (ILs) provides one means to enhance drug solubility in LBFs. However, whether different ILs benefit from formulation in different LBFs is largely unknown. In the current studies, lumefantrine was isolated as a number of different lipophilic salt/ionic liquid forms and performance was assessed after formulation in a range of LBFs. The solubility of lumefantrine in LBF was enhanced 2- to 80-fold by isolation as the lumefantrine docusate IL when compared to lumefantrine free base. The increase in drug loading subsequently enhanced concentrations in the aqueous phase of model intestinal fluids during in vitro dispersion and digestion testing of the LBF. To assess in vivo performance, the systemic exposure of lumefantrine docusate after administration in Type II-MCF, IIIB-MCF, IIIB-LCF, and IV formulations was evaluated after oral administration to rats. In vivo exposure was compared to control lipid and aqueous suspension formulations of lumefantrine free base. Lumefantrine docusate in the Type IIIB-LCF showed significantly higher plasma exposure compared to all other formulations (up to 35-fold higher). The data suggest that isolation of a lipid-soluble IL, coupled with an appropriate formulation, is a viable means to increase drug dose in an oral formulation and to enhance exposure of lumefantrine in vivo.

Keywords: ionic liquid; lipophilic salt; drug delivery; drug absorption; poorly water-soluble drug; lumefantrine; lipid-based formulation; SEDDS; lipid formulation classification system

---

1. Introduction

Many currently marketed drugs, and drugs in development, are poorly water-soluble and classified as class II or class IV as defined by the Biopharmaceutics Classification System (BCS) [1–4]. Class II compounds are poorly water-soluble, but have high membrane permeability, while class IV compounds have both poor water solubility and poor permeability. Reformulation approaches are commonly employed to improve the bioavailability of class II compounds; however, for class IV compounds, the challenge is significantly greater and whilst improvements are possible via formulation approaches, it is often more beneficial to develop alternative analogues with more suitable physicochemical properties [5].

The increasing prevalence of poorly water-soluble drugs has necessitated the development of a range of formulation approaches to increase apparent drug solubility in the gastro-intestinal fluids [6–9]. These include the use of solid dispersions, lipid-based formulations, cyclodextrins, surfactants, and particle size reduction techniques [7]. The current studies have focused on the use of lipid-based formulations in conjunction with alternate lipophilic salt forms (or ionic liquids) to promote drug solubility in lipid-based formulations. The anti-malarial drug, lumefantrine, has been employed as a model poorly water-soluble drug. Lumefantrine is currently formulated as a tablet and is dosed in combination with artemether [10]. The oral bioavailability of lumefantrine is 4–11% [11]. The low and likely solubility limited bioavailability of lumefantrine may be improved by administration with a fatty meal, but this leads to variable bioavailability [11].

Ionic liquids (ILs) are generally defined as organic salts with melting points below 100 °C [12–14]. They are usually composed of an organic cation and an inorganic/organic anion [15,16]. Inefficient crystal packing of the ions and/or a more diffuse charge on either or both ionic species leads to the weakening of the interactions between the cation and anion in ILs. Using a bulky or irregularly shaped counterion further decreases inter-molecular interactions and the efficiency of crystal packing, leading to a decreased melting point [1,17,18]. The reduction in melting point observed in ILs typically increases solubility in both aqueous and non-aqueous vehicles, but has been employed here to enhance solubility in non-aqueous LBF vehicles.

Active pharmaceutical ingredient-ionic liquids (API-ILs) comprise a drug and an appropriate counterion. For the formation of an IL, there are a wide range of cation-anion combinations available and previous studies have shown that different API-ILs can result in improvements in drug solubility, stability, bioavailability, and membrane permeability [1,19–23]. The flexibility of counterion choice enables selection of an appropriate counterion for specific drug delivery systems. Here, we focus on lipophilic counterions to increase the solubility of ILs in lipid-based formulations.

Lipid-based formulations (LBFs) are composed of combinations of traditional lipids (such as monoglycerides, diglycerides, and triglycerides), surfactants, and co-solvents [24,25]. LBFs are commonly used as vehicles for poorly water-soluble drugs as they increase drug solubilization in the gastro-intestinal (GI) tract via integration into endogenous lipid solubilization pathways. As the drug is usually pre-dissolved in the formulation, LBFs also avoid traditional solid–liquid dissolution in the GI tract, a process that is often the rate-limiting step in absorption [26–28].

The solubilization capacity of LBFs, however, typically changes as the formulation is digested and interacts with bile salt micelles in the GI tract, leading to changes in structure [5]. Drug solubilization is more efficient at high lipid loads, especially in the presence of medium-chain lipids. These medium-chain lipids are digested very rapidly and efficiently, and at high concentration, promote drug solubilization [27,29]. However, at lower lipid loads, medium-chain lipids can lead to drug precipitation and reduced absorption as they are digested to form medium-chain fatty acids. Medium-chain fatty acids are relatively polar, and at low concentrations, swell bile salt micelles relatively poorly (thereby reducing overall solubilization capacity). In contrast, long-chain lipid digestion products are less polar, swell bile salt micelles more effectively, and typically lead to more robust drug solubilization, even at low lipid concentrations [28].

In 2006, Pouton developed a general system for the classification of lipid-based formulations (the Lipid Formulation Classification System (LFCS)). The LFCS aids in identifying the critical performance characteristics of lipid systems [5]. Type I formulations require digestion and are composed of only oils. Type II formulations contain lipids and water-insoluble surfactants and self-emulsify on contact with the GI fluids (i.e., self-emulsifying drug delivery systems (SEDDS)) [30–32]. Type III formulations are self-microemulsifying drug delivery systems (SMEDDS) or self-nanoemulsifying drug delivery systems (SNEDDS) and are sub-classified into Type IIIA and Type IIIB. Type IIIA formulations contain lipids, water-soluble surfactants and/or co-solvents, and Type IIIB contain similar components but have greater proportions of water-soluble surfactants and co-solvents than Type IIIA. Type IV formulations contain only hydrophilic surfactants and co-solvents [5].

A potential limitation of the utility of LBFs is low drug loading in the formulation such that the therapeutic drug dose cannot be solubilized in a volume of lipid that can be filled into one or two capsules [6]. The use of ionic liquid forms of drugs can increase drug loading in LBFs [6,32–34], but whether solubilization is maintained at these higher drug loads as the formulation is dispersed and digested is not well understood. Similarly, the relationship between formulation type and composition and the fate of an API-IL during formulation digestion has not been widely explored. The aim of the current study was to explore the ability of IL technology to promote lumefantrine solubility in range of LFCS class formulations, to explore the ability of the high load LBF to maintain drug in a solubilized state during dispersion and digestion, and ultimately to evaluate whether this translates into improved systemic exposure after oral administration. Recent interest in the ability of LBF to promote drug supersaturation in the GI tract also promoted an analysis as to whether the high drug loads that IL formation facilitate also allow for the facile generation of highly supersaturated conditions in the GI tract.

2. Materials and Methods

Materials. Lumefantrine was purchased from AK Scientific (Union City, CA, USA). Dioctyl sulfosuccinate sodium salt (sodium docusate) was purchased from Sigma Aldrich (St. Louis, MO, USA). Sodium dodecyl sulfate was purchased from BASF (Castle Hill, NSW, Australia). Hydrogen chloride in diethyl ether (2.0 M) was obtained from Sigma-Aldrich (St. Louis, MO, USA). Chloroform, methanol, and dichloromethane were purchased from Merck (Melbourne, VIC, Australia). Details of the LBFs used in the study are provided in Table 1. Captex® 355 EP/NF and Capmul® MCM EP were supplied by Anzchem (Sydney, NSW, Australia). Tween™ 85 was supplied by Croda (Sydney, NSW, Australia). Kolliphor® RH 40 was obtained from BASF (Melbourne, VIC, Australia). Maisine™ 35-1 was supplied by Trapeze Associates Pty. Ltd. (Sydney, NSW, Australia). Soybean oil and butylated hydroxytoluene (BHT) were purchased from Sigma Aldrich (St. Louis, MO, USA). Ethanol was purchased from Merck (Melbourne, VIC, Australia). Sodium taurodeoxycholate > 95% (NaTDC), 4-bromophenylboronic acid, and porcine pancreatin (8 X USP specification activity) were purchased from Sigma Aldrich (St. Louis, MO, USA). Phosphatidylcholine (PC) (Lipoid E PC S, ~99.2% pure, from egg yolk) was obtained from Lipoid (Ludwigshafen, Germany). The 0.6 M and 0.2 M strength sodium hydroxide solutions were diluted from a stock solution of 1.0 M sodium hydroxide that was purchased from Science Supply (Melbourne, VIC, Australia). Formic acid was purchased from Sigma Aldrich (Sydney, NSW, Australia) and acetonitrile, both HPLC and LC-MS grade, was purchased from Merck (Melbourne, VIC, Australia). All solvents used were of analytical purity or high-performance liquid chromatography (HPLC) grade.

Table 1. Composition of the lipid-based formulations used for equilibrium solubility and in vitro dispersion and digestion studies. All formulations contained ~1% butylated hydroxytoluene as an anti-oxidant.

Formulation Type	Component	Composition (% *w/w*)
I-MC	Captex® 355 EP/NF (Glyceryl tricaprylate/caprate)	50.0
	Capmul® MCM EP (Glyceryl monocaprylocaprate)	50.0
II-MC	Captex® 355 EP/NF (Glyceryl tricaprylate/caprate)	32.5
	Capmul® MCM-EP (Glyceryl monocaprylocaprate)	32.5
	Tween™ 85 (Polyoxyethylene sorbitan trioleate)	35.0
IIIA-MC	Captex® 355 EP/NF (Glyceryl tricaprylate/caprate)	32.5
	Capmul® MCM EP (Glyceryl monocaprylocaprate)	32.5
	Kolliphor® RH 40 (Polyoxyl 35 castor oil)	35.0
IIIB-MC	Capmul® MCM-EP (Glyceryl monocaprylocaprate)	25.0
	Kolliphor® RH 40 (Polyoxyl 35 castor oil)	50.0
	ethanol	25.0
I-LC	Maisine™ 35-1 (Glyceryl monolinoleate)	50.0
	soybean oil	50.0
II-LC	Maisine™ 35-1 (Glyceryl monolinoleate)	32.5
	soybean oil	32.5
	Tween™ 85 (Polyoxyethylene sorbitan trioleate)	35.0
IIIA-LC	Maisine™ 35-1 (Glyceryl monolinoleate)	32.5
	soybean oil	32.5
	Kolliphor® RH 40 (Poloxyl 40 hydrogenated castor oil)	35.0
IIIB-LC	Maisine™ 35-1 (Glyceryl monolinoleate)	25.0
	Kolliphor® RH 40 (Poloxyl 40 hydrogenated castor oil)	50.0
	ethanol	25.0
IV	Kolliphor® RH 40 (Poloxyl 40 hydrogenated castor oil)	50.0
	ethanol	50.0

Lumefantrine Hydrochloride Preparation. Lumefantrine (3 g, 5.66 mmol) was dissolved in anhydrous dichloromethane (50 mL) and a solution of 2.0 M hydrochloride in diethyl ether was added dropwise to the solution (2.85 mL, 5.66 mmol). The solution was stirred for 3 h at room temperature. Dichloromethane was then evaporated and the residual solid product was dried under vacuum. (Mass: 3.02 g, Yield: 94%).

Ionic Liquid Preparation: Lumefantrine Dodecyl Sulfate. Lumefantrine hydrochloride (1.5 g, 2.65 mmol) was dissolved in methanol (200 mL) and sodium dodecyl sulfate (0.77 g, 2.65 mmol) was added. The solution was stirred for 3 h at room temperature. The methanol was then evaporated in vacuo and the residue was dried under vacuum overnight. The residue was treated with chloroform (200 mL) to dissolve the IL and precipitate out the sodium chloride. The mixture was filtered through a 2 μm micro filter, the solvent evaporated, and the final solid product dried under vacuum. (Mass: 1.76 g, Yield: 83%).

Ionic Liquid Preparation: Lumefantrine Docusate. Lumefantrine hydrochloride (1.5 g, 2.65 mmol) was dissolved in methanol (200 mL) and dioctylsulfosuccinate sodium salt (docusate) (1.2 g, 2.65 mmol) was added. The solution was stirred for 3 h at room temperature. The methanol was then evaporated in vacuo and the residue was dried under vacuum overnight. The residue was treated with chloroform (200 mL) to dissolve the IL and precipitate out the sodium chloride. The mixture was filtered through a 2 μm micro filter, the solvent evaporated, and the final semi-solid product dried under vacuum. (Mass: 1.96 g, Yield: 77%).

Nuclear Magnetic Resonance (NMR) Spectroscopy. ^{1}H and ^{13}C NMR spectra were obtained at 400.13 Hz and 100.62 Hz respectively, on a Bruker Advance III Nanobay 400 MHz spectrometer coupled to the BACS 60 automatic sample changer. All spectra were processed using MestReNova 6.0 software. The chemical shifts of all ^{1}H signals were measured relative to the expected solvent peaks of the NMR solvent; 2.50 ppm (DMSO-d_6). The data for all spectra are reported in the following format: chemical shift (integration, coupling constant J (Hz), multiplicity). Multiplicity is defined as; s = singlet, d = doublet, t = triplet, q = quartet, dd = doublet of doublets, dt = doublet of triplets, and m = multiplet. Subsequent abbreviations also include J (Hz) = coupling constant in Hertz.

The ^{1}H and ^{13}C NMR spectra and MS data for lumefantrine hydrochloride, lumefantrine dodecyl sulfate, and lumefantrine docusate are provided in the supporting information.

Liquid Chromatography-Mass Spectroscopy (LC-MS). LC-MS chromatograms were obtained using an Agilent 6100 Series Single Quad LC/MS coupled with an Agilent 1200 Series HPLC, 1200 Series G1311A quaternary pump, 1200 series G1329A thermostatted autosampler, and 1200 series G1314B variable wavelength detector. The conditions for liquid chromatography were: reverse phase HPLC analysis using a Phenomenex Luna C_8(2) 5 μm (50 × 4.6 mm) 100 Å column at a temperature of 30 °C. 5 μL of sample was injected and the sample was run in solvent A of 99.9% acetonitrile, 0.1% formic acid with a gradient of 5–100% (*v*/*v*) solvent A over 10 min. Solvent B was 99.9% water with 0.1% formic acid. Detection was at a UV wavelength of 254 nm. The conditions for mass spectrometry were: quadrupole ion source with multimode-ES, drying gas temperature 300 °C, and vaporizer temperature 200 °C. The capillary voltage was 2000 V in positive mode, or 4000 V in negative mode and the scan range was 100–1000 *m*/*z* with a step size of 0.1 s over 10 min.

High-Resolution Mass Spectrometry. All high-resolution mass spectrometry analyses were performed on an Agilent 6224 TOF LC/MS Mass Spectrometer coupled to an Agilent 1290 Infinity HPLC (Agilent, Palo Alto, CA, USA). All data were acquired and reference mass corrected via a dual-spray electrospray ionization (ESI) source. Each scan or data point on the Total Ion Chromatogram (TIC) is an average of 13,700 transients, producing a spectrum every second. Mass spectra were created by averaging the scans across each peak and background subtracting against the first 10 s of the TIC. Acquisition was performed using the Agilent Mass Hunter Data Acquisition software version B.05.00 Build 5.0.5042.2 and analysis was performed using Mass Hunter Qualitative Analysis version B.05.00 Build 5.0.519.13. The MS conditions were: electrospray ionization, a drying gas flow of 11 L/min at a temperature of 325 °C, a nebulizer at 45 psi, a capillary voltage of 4000 V, the fragmentor at 160 V, the skimmer at 65 V, and the OCT RFV of 750 V. The scan range acquired was 100–1500 *m*/*z*. The internal reference ions in positive ion mode had a *m*/*z* of 121.0509 and 922.0098. Chromatographic separation was performed using an Agilent Zorbax SB-C18 Rapid Resolution HT 2.1 × 50 mm, 1.8 μm column (Agilent Technologies, Palo Alto, CA, USA) using an acetonitrile gradient (5% to 100%) over 3.5 min at 0.5 mL/min.

Polarized Light Microscopy and Hot Stage Microscopy. The melting point ranges of all compounds were assessed using a hot stage microscope on an Axiolab Laboratory Microscope (manufactured 1997, S/N 982650) supplied by Carl Zeiss (Carl Zeiss, Oberkochen, Germany). The microscope was fitted with cross polarizing filters and coupled to a Linkam HFS91 hot stage connected to a Linkam TP93 system controller (Linkam Scientific Instruments, Tadworth, UK). Samples, mounted between two glass coverslips, were heated at 5° C/min until the compound showed signs of melting, at which time the rate was decreased to 1° C/min, and this was monitored continuously. Images were captured with a Canon LA-DC52C PowerShot A70 camera at 10 × magnification using Canon Utilities RemoteCapture version 2.7.2.16 software. Complete melting was defined as the lowest temperature at which the sample was free of birefringence; in the case of the amorphous sample (lumefantrine docusate), complete melting was defined as the lowest temperature where no solid structures were evident and therefore might also be described as a solid–liquid transition temperature.

Equilibrium Solubility Studies. The equilibrium solubility of lumefantrine, lumefantrine hydrochloride, lumefantrine dodecyl sulfate, and lumefantrine docusate in each formulation was determined by initially adding 20 mg of compound to 200 mg of LBF in a microcentrifuge tube. Each equilibrium solubility experiment was completed in triplicate. Equilibrium solubility was defined when solubility measured across two consecutive days varied less than 5%. The samples were allowed to equilibrate at 37 °C and were vortex-mixed twice a day to ensure the compounds were well dispersed in the LBF. The samples were left to equilibrate for at least 3 days. If all the compound was observed to be fully dissolved, the sample was centrifuged for 15 min at 14,800 rpm (21,000× *g*) at 37 °C (Thermo Scientific, Heraeus Pico 21 Centrifuge, Langenselbold, Germany) to confirm complete dissolution. Where no pellet was observed, another 20 mg of compound was added, and the process was repeated until complete dissolution was no longer observed and excess IL was evident. At each time point, samples were then centrifuged at 37 °C (14,800 rpm (21,000× *g*), Thermo Scientific, Heraeus Pico 21 Centrifuge) and aliquots of 20 mg were taken from the supernatant and dissolved in 1 mL of chloroform: methanol (2:1, *v*/*v*). The chloroform: methanol solution was then diluted 20-fold with water: acetonitrile (1:1, *v*/*v*). The samples were then assayed by HPLC to determine the concentration of compound in formulation.

In Vitro Drug Solubilization. The LBF were assayed using previously reported in vitro dispersion and digestion tests in order to assess the potential for formulations containing dissolved API-IL to maintain drug in a solubilized state as the formulation is dispersed and digested under simulated gastro-intestinal conditions [35]. Drug loading in each formulation was at 80% of the equilibrium solubility of the drug or API-IL in that formulation to maintain a consistent thermodynamic activity across all LBFs.

In vitro dispersion and digestion experiments were conducted using a pH-stat apparatus (Metrohm® AG, Herisau, Switzerland), which comprised a Titrando 802 propeller stirrer/804 Ti Stand combination, a glass pH electrode (iUnitrode), and two 800 Dosino dosing units coupled to 10 mL autoburettes (Metrohm® AG). The pH-stat was connected to a PC and operated via Tiamo 2.0 software (Metrohm®) [35]. First, 1.100 g of formulation was dispersed in 40 mL of bile salt/phospholipid micelles in simulated intestinal fluid (2 mM tris-maleate, 150 mM NaCl, 1.4 mM $CaCl_2{\cdot}2H_2O$, 3 mM NaTDC, 0.75 mM PC, pH 6.5, 37 °C) and then 4 mL of pancreatic enzyme extract (pancreatin) was added to stimulate digestion. The pancreatic enzyme was prepared by mixing 1 g of porcine pancreatin with 5 mL of lipolysis buffer and 20 μL of 5 M NaOH. After mixing, this was then centrifuged at 2880× *g* at 5 °C for 10 min (Crown Scientific, Eppendorf AG Centrifuge 5804 R, Sydney Australia) and the supernatant used as the enzyme solution. The lipolysis buffer was prepared by dissolving 0.474 g of tris maleate, 0.206 g of $CaCl_2{\cdot}H_2O$, and 8.775 g of NaCl in one liter of distilled water to form a 2 mM tris maleate 1.4 mM $CaCl_2{\cdot}H_2O$ 150 mM NaCl buffer; the pH was adjusted to 6.5 using NaOH. The micellar solution was prepared by dissolving 0.783 g of NaTDC and 0.291 g of PC in 500 mL of lipolysis buffer to form a 3 mM NaTDC 0.75 mM PC solution. The temperature-controlled vessel was held at 37 °C, and contained bile salt/phospholipid micellar solution in simulated intestinal fluid.

After dispersion of the formulation, digestion was initiated by addition of the pancreatic lipase/co-lipase solution. This resulted in the liberation of fatty acids (FAs) and therefore a drop in pH. This drop in pH was detected by the pH-stat controller, which then titrated the FA produced using an autoburette that added NaOH to keep the pH at a set point (0.6 M NaOH for MCFs, and 0.2 M

NaOH for LCFs). By knowing the quantity of NaOH added and assuming stoichiometric titration, this indirectly quantified the extent of FA production (as a measure of digestion). At time intervals of −10, −5, and 0 (i.e., in the 15 min dispersion phase), and 5, 10, 15, 30, and 60 min (digestion phase—i.e., after initiation of digestion at time = 0), 1 mL aliquots were taken. Samples taken during digestion were immediately treated with a lipolysis inhibitor (10 μL of 0.5 M 4-bromophenylboronic acid in methanol per 1 mL of digestion medium) to halt digestion. All samples were then centrifuged to form a maximum of 3 phases—an oil phase, an aqueous phase, and a pellet phase.

For the Type I-LCF and II-LCF formulations (where an oil phase was present), samples were separated using an ultracentrifuge (4 mL sample volume, 55,000 rpm at 37 °C for 30 min; Optima™ XE-90 Ultracentrifuge, SW60Ti rotor; Beckman Coulter, Palo Alto, CA, USA). For all other formulations, a Thermo Scientific, Heraeus Pico 21 Centrifuge was used (1 mL sample volume, 15 min at 21,000× *g*, 37 °C). The mass of the drug recovered in the aqueous phase and the pellet phase was then quantified by HPLC [35,36]. The in vitro digestion apparatus is shown in Figure 1. The supernatant and pellet samples were diluted with acetonitrile (1:10 *v*/*v*) and then mobile phase (1:10 *v*/*v*), and lumefantrine concentrations in the samples determined using HPLC [35–38].

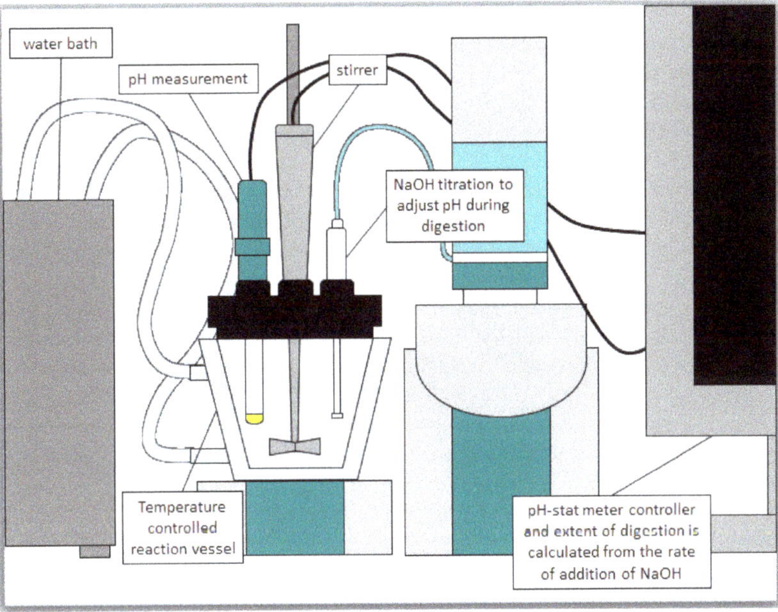

Figure 1. In vitro digestion model for testing lipid formulations. A temperature controlled (37 °C) vessel containing digestion buffer, bile salt, and phospholipid is used. The lipid-based formulations are added to the vessel, and digestion commences when pancreatic lipase and co-lipase are added. When digestion begins, fatty acids are liberated, which causes a transient drop in pH. A pH electrode coupled to a pH-stat controller and autoburette quantifies the drop in pH, and automatically titrates the liberated fatty acids by adding an equimolar quantity of sodium hydroxide. This maintains the pH at a set point and facilitates indirect quantification of the extent of digestion. Samples are taken over time and centrifuged to separate the digest into an oil phase, an aqueous phase, and a pellet phase, which are then analyzed for drug content [28].

Supersaturation During In Vitro Testing. The degree of supersaturation of drug concentrations in the aqueous phase during formulation dispersion/digestion was assessed by comparing the measured aqueous phase concentration to equilibrium drug solubility in blank aqueous phase obtained by dispersion and digestion of blank (i.e., non-drug loaded) formulation under identical conditions [39].

Due to the physical state of lumefantrine docusate (i.e., an amorphous semi-solid), the equilibrium solubility assessment was conducted by dissolving 3 mg of IL in 10 μL of ethanol in a microcentrifuge tube and then adding 200 mg of aqueous phase obtained from the blank digestions (n = 3 for each time point for each formulation). The contents were subsequently vortex-mixed for 30 s. Samples were collected at 0.5, 1, 2, 3, and 4 h after addition of blank aqueous phase and then centrifuged (10 min at 14,800 rpm (21,000× *g*) at 37 °C (Thermo Scientific, Heraeus Pico 21 Centrifuge) and assayed by HPLC to determine the drug concentration. Equilibrium solubility was defined when the concentration measured differed by less than 5% across sequential samples. The supersaturation ratio was then calculated as the ratio of the drug concentration measured in the aqueous phase during the dispersion/digestion test to the equilibrium solubility of drug in the aqueous phase at that time point. Supersaturation was assessed at 4 time points during the in vitro dispersion and digestion tests (0 min (end dispersion), and 5, 30, and 60 min (post digestion)) for the Type II-MC, IIIB-MC, IIIB-LC, and IV formulations.

HPLC Assay Conditions for Lumefantrine. All HPLC analyses were conducted using a Waters Alliance 2695 Separation Module (Waters Alliance Instruments, Milford, CT, USA). The column was a reverse phase C-18 Phenomenex 3 μm, 100 × 4.6 mm column. The injection volume was 50 μL and UV detection was at 254 nm. The chromatography was run isocratically. The mobile phase consisted of water with 0.1% *w*/*v* formic acid (mobile phase A) and acetonitrile with 0.1% *w*/*v* formic acid (mobile phase B). For quantification of the equilibrium solubility samples, the ratio was fixed at 45:55 *v*/*v* (A/B) while for the in vitro dispersion and digestion samples, the ratio was 50:50 *v*/*v* (A/B). The flow rate was 1 mL/min and the retention times were ~2.5 and 5.3 min respectively. The calibration standards were 50, 200, 1000, 2000, 5000, 10,000, and 20,000 ng/mL.

Validation of the lumefantrine HPLC assay was run over two days. Intra-assay accuracy was determined by replicated analysis (n = 5) of three standards at the lowest, middle, and highest concentrations (50, 5000, and 20,000 ng/mL). Inter-assay accuracy was determined on two separate days. The data were expressed as a percentage of the measured concentration over the theoretical concentration. The mean accuracy of the lowest concentration (50 ng/mL) was within ±15% of the theoretical concentration, while the mean accuracy of the middle and highest concentrations (5000 and 20,000 ng/mL) was within ±10% of the theoretical concentration. Intra-assay precision and inter-assay precision were calculated in both runs for each of the three concentrations and expressed as the coefficient of variation. Precision was within ±10% for all three concentrations. Linearity was performed on the standard curves for each analysis and linearity was accepted when the correlation coefficient (r^2) of the regression line was >0.99.

Oral Bioavailability Studies. All procedures were approved by the Monash Institute of Pharmaceutical Sciences Animal Ethics Committee (Approval code 13227. Approval date: 10 April 2014). Experiments were conducted in fasted male Sprague-Dawley rats (240–320 g). The day prior to the study, the rats were anaesthetized with isoflurane and the right carotid artery was surgically cannulated with polyethylene tubing to facilitate blood collection (procedure described previously) [40]. Animals were allowed to recover overnight and were fasted up to 12 h prior to and 8 h after dose administration, with water provided ad libitum. The rats were fed 8 h after dosing. The general approach to the doses chosen for the in vivo studies was based on the desire to provide evidence of utility of the IL approach and to provide a comparison between solution and suspension formulations. In general, the potential drug load that could be dissolved using the IL was much higher than that that could be achieved with the free base, and so to compare exposure at a fixed dose (and typically a relatively high dose since this was possible with the IL technology), the IL formulations were solutions whereas the free base formulations were suspensions. However, data were also collected (where possible) at a fixed dose when both free base and IL form were in solution. These studies had to be conducted at a lower dose to allow the free base to be in solution. The compounds used were employed as model low aqueous solubility drugs and therefore the absolute doses chosen were not chosen to have any pharmacological relevance.

The LBFs containing lumefantrine docusate were dispersed (250 mg LBF in 1 mL of water) immediately prior to oral gavage to lightly anaesthetized rats. Animals were then administered a further 0.5 mL of water. The lumefantrine docusate was dissolved in Type II-MCF, Type IIIB-MCF, Type IIIB-LCF, and Type IV formulations to provide a dose of 50, 65, 65, or 85 mg/kg lumefantrine free base, respectively. These doses were equivalent to 80% of equilibrium solubility of lumefantrine docusate in the respective LBFs. Lumefantrine free base was administered as a suspension in the Type IIIB-LCF (i.e., what was expected to be the most effective LBF) at a dose of 85 mg/kg, and dosed as an aqueous suspension at 85 mg/kg. The aqueous suspension comprised 0.5% *w*/*v* sodium carboxymethylcellulose, 0.4% *w*/*v* Tween 80, and 0.9% NaCl in water.

Blood samples were collected via the carotid artery cannula at pre-dose, 0.5, 1, 2, 3, 4, 6, 8, 10, and 24 h after oral administration of the formulations. Blood samples were centrifuged at 6700× *g* for 5 min and plasma was collected and stored at −20 °C until assayed by LC-MS. Statistically significant differences between formulations were determined by one-way analysis of variance (ANOVA) at a test significance level of $\alpha \leq 0.05$, followed by a Tukey's multiple comparisons test.

Plasma Sample Preparation and Analysis. Lumefantrine concentrations in rat plasma were assayed via LC-MS using a modification of a previously published assay for lumefantrine [41]. Calibration standards of lumefantrine were prepared by spiking blank rat plasma (50 µL) with lumefantrine standard solutions (10 µL) in 1:1 *v*/*v* water: acetonitrile to give plasma concentrations in the range of 50 to 5000 ng/mL. Plasma samples or calibration standards (50 µL) were spiked with 10 µL of internal standard (halofantrine hydrochloride, 1000 ng/mL in 1:1 *v*/*v* water: acetonitrile) followed by vortex mixing for 10 s. To precipitate plasma proteins, acetonitrile (200 µL) was added and samples vortex mixed for 1 min. The samples were then allowed to stand at room temperature for 30 min. After centrifugation at 10,000× *g* for 10 min at room temperature, 150 µL of supernatant was transferred into vials for analysis.

Lumefantrine plasma samples were analyzed using a Shimadzu LCMS-8050 system (Shimadzu Scientific Instruments, Kyoto, Japan) consisting of a CBM-20A system controller, a DGU-20A5R degassing unit, two Nexera X2 LC-30 AD liquid chromatograph pumps, a Nexera X2 SIL-30AC autosampler, a CTO-20A column oven (held at 40 °C), and a triple quadrupole mass spectrometer with an electrospray ionization (ESI) interface. Data acquisition and processing was performed using LC-MS Solutions software (Shimadzu, Kyoto, Japan). The desolvation line (DL) and heat block were kept at 250 °C and 400 °C, respectively. The nebulizing gas flow and drying gas flow rates were 3.0 L/min and 10.0 L/min, respectively. Mobile phase A was Milli-Q water containing 0.1% *w*/*v* formic acid, and mobile phase B was acetonitrile containing 0.1% *w*/*v* formic acid. The mobile phase flow rate was 0.5 mL/min with the following gradient elution: mobile phase B was first held at 30% for 1.25 min, then linearly increased to 80% over the next 0.75 min. Mobile phase B was then held at 80% for 0.5 min, followed by a linear decrease to 30% over the next 0.5 min after which conditions were held for another 0.5 min. Each sample (1 µL) was injected onto a Phenomenex Kinetex® C18 column (2.6 µm, 100 Å, 50 × 2.1 mm, Sydney, NSW, Australia). The retention times of lumefantrine and the internal standard (halofantrine hydrochloride) were 2.1 and 1.9 min, respectively. Lumefantrine detection was achieved using positive electrospray ionization with multiple-reaction monitoring (MRM) of the 530.2 > 512.2 mass/charge ion peak (*m*/*z*) at a collision energy of −25.0 V. The internal standard halofantrine hydrochloride was monitored at 500.2 > 142.0 *m*/*z* at a collision energy of −25.0 V. Sample concentrations were determined by comparison to a calibration curve obtained by fitting the peak area ratio of lumefantrine to internal standard (halofantrine hydrochloride) versus concentration data to a linear equation with a weighting factor inversely proportional to the standard concentration.

Validation of the lumefantrine LC-MS plasma assay was run over two days. Intra-assay accuracy was determined by replicate analysis ($n = 5$) of three standards at the lowest, middle, and highest concentrations (50, 2000, 5000 ng/mL). Inter-assay accuracy was determined on two separate days. The data were expressed as a percentage of the measured concentration over the theoretical concentration. The mean accuracy was within ±10% of the theoretical concentration. Intra-assay precision and inter-assay precision were calculated for both runs for each of the three concentrations and expressed as the coefficient of variation. Precision was within ±10% for all three concentrations. Linearity was performed on the standard curves for each analysis and linearity was accepted when the correlation coefficient (r^2) of the regression line was >0.99.

3. Results

3.1. Characterization of Lumefantrine Compounds

The measured melting temperatures of the lumefantrine ILs and related compounds are listed in Table 2. Detailed NMR spectroscopy data are listed in the experimental section. The melting point of the lumefantrine ILs appeared to decrease with increasing bulk/complexity of the counterion. Lumefantrine hydrochloride is a yellow powder and shows birefringence under polarized light microscopy, indicating crystallinity. As such, it has the highest melting point (180–200 °C). Lumefantrine free base and lumefantrine dodecyl sulfate are also yellow powders which show some degree of birefringence under the cross-polarized light microscope, but have lower melting points than the hydrochloride salt (133–140 °C and 120–128 °C, respectively). Since the melting point of lumefantrine dodecyl sulfate was >100 °C, it is more appropriately termed a lipophilic salt rather than an ionic liquid [33]. As expected, the complexity of the docusate counterion resulted in isolation of lumefantrine docusate as a clear yellow semi-solid with a much lower melting range of 52–60 °C. Several attempts were made to recrystallize lumefantrine docusate but a crystal form could not be isolated.

Table 2. The melting point range and appearance under cross-polarized light of lumefantrine free base, the traditional salt lumefantrine hydrochloride and the lipophilic salts lumefantrine dodecyl sulfate and lumefantrine docusate. The lower melting range (<100 °C) of lumefantrine docusate further classifies this material as an ionic liquid.

Compound	Cross-Polarized Light Microscope Image	Melting Point Range (°C)
lumefantrine free base	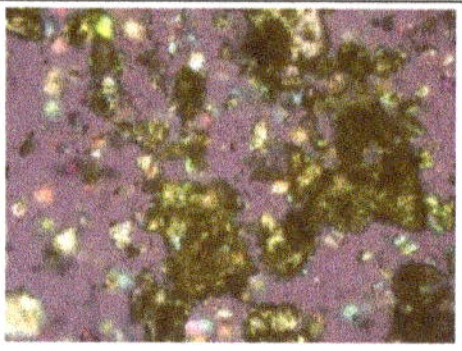	133–140
lumefantrine hydrochloride	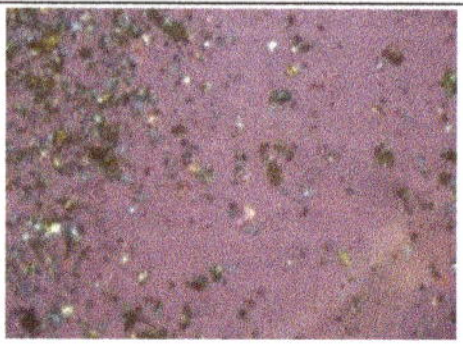	180–200
lumefantrine dodecyl sulfate	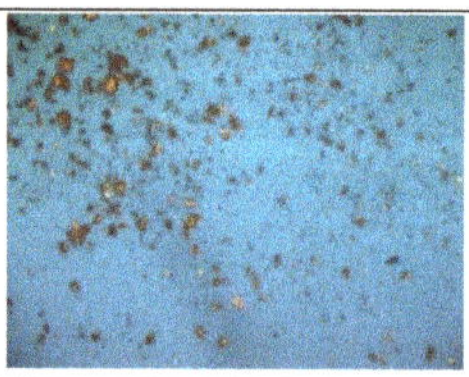	120–128
lumefantrine docusate	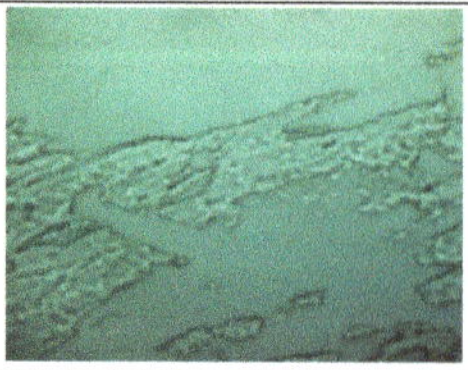	52–60

3.2. Lumefantrine Equilibrium Solubility in Lipid-Based Formulations

The equilibrium solubility of the lumefantrine compounds was assessed in each of the LBFs described in Table 1. The equilibrium solubility of lumefantrine, lumefantrine hydrochloride, lumefantrine dodecyl sulfate, and lumefantrine docusate in the different LBFs is shown in Figure 2, where the data are presented as the equivalent concentration of free base. In general, lumefantrine solubility was higher in the medium-chain formulations than in the long-chain formulations, and solubility was higher in the more polar formulations such as the Type IIIB and Type IV formulation. This is consistent with previous work, with both non-IL drugs and ILs [32,35]. Isolation of lumefantrine as the dodecyl sulfate IL resulted in increases in solubility in the Type IIIB and Type IV formulations, but did not result in significant advantages in lipid solubility in the others. In contrast, isolation as the docusate IL markedly enhanced solubility in all LBF (2-to-80-fold higher than lumefantrine, depending on formulation) [35].

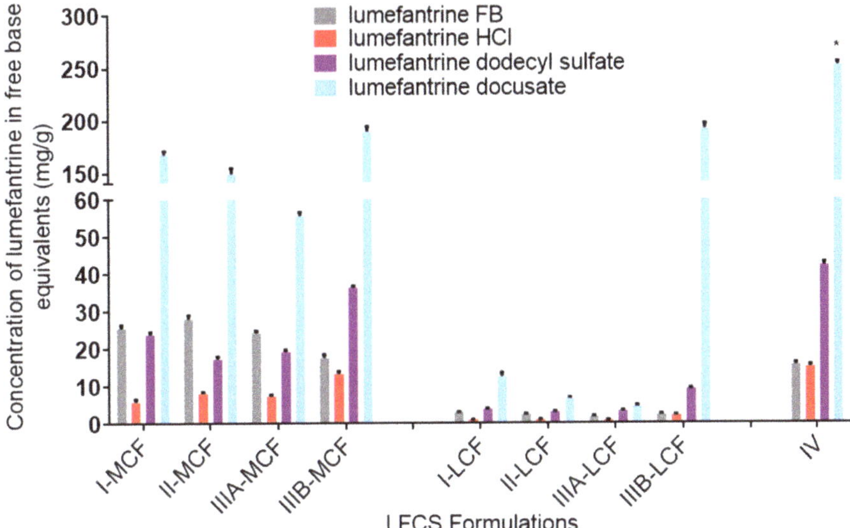

Figure 2. Equilibrium solubility of lumefantrine compounds in lipid-based formulations. Data are expressed in lumefantrine free base equivalents. Data are mean ± SD, $n = 3$. * Lumefantrine docusate was miscible in the Type IV formulation and the data shown reflect the measured concentration after mixing the docusate and formulation in a 1:1 *w*/*w* ratio.

3.3. In Vitro Evaluation of Lumefantrine and Lumefantrine Docusate in LBF

In light of the lower improvement in lipid solubility for the dodecyl sulfate IL, in vitro dispersion and digestion tests were conducted for LBF containing lumefantrine free base and lumefantrine docusate only. The drug-phase distributions for lumefantrine and lumefantrine docusate are shown in Figures 3 and 4, respectively. A comparison of the aqueous phase concentrations attained for LBF containing lumefantrine and lumefantrine docusate (at loading levels of 80% of equilibrium solubility) at the end of the dispersion and digestion phase of experiments is provided in Figures 5 and 6, respectively.

After centrifugation, samples were separated into an oil phase (if present), an aqueous phase, and a pellet phase (all individual in vitro dispersion and digestion data are provided in the Supporting Information). For both lumefantrine and lumefantrine docusate, an oil phase was present more commonly after dispersion and digestion of the long chain lipid-containing formulations (LCF) and was only present for the Type I medium chain formulation (MCF). The Type I formulations contain only lipids and no surfactant and as such disperse poorly resulting in phase separation. Initiation of digestion, however, results in the generation of more amphiphilic digestion products that promote dispersion and recovery in the aqueous phase and reduce drug recovery in the oil phase. Dispersion of the Type II-MCF resulted in the recovery of a large proportion of the drug in the 'pellet' phase; however, this likely reflects phase separated (dense) lipid rather than drug precipitation [42]. This suggestion is consistent with a significant reduction in drug recovery in the pellet after digestion is initiated (where digestion results in a change to the nature of the oil phase, an increase in amphiphilicity, and enhancement in dispersion). For both long and medium chain formulations, the Type IIIA and IIIB formulations resulted in good solubilization after both dispersion and digestion and the majority of the drug was present in the aqueous phase. The only significant differences in performance with respect to formulation class were the Type I-MCF, Type II-MCF and Type IV formulations, where significantly more drug precipitation was apparent for the lumefantrine docusate formulation when compared to lumefantrine. This likely reflects the much higher drug loading afforded by the use of the docusate IL.

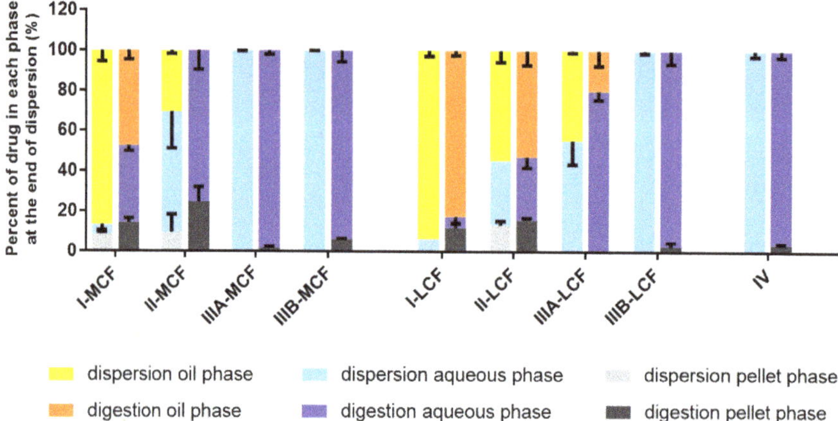

Figure 3. Drug distribution of lumefantrine during in vitro dispersion and digestion tests of LBFs. Data presents % drug in each phase at the end of dispersion phase and at the end of digestion phase. Data are mean ± SD, $n = 3$.

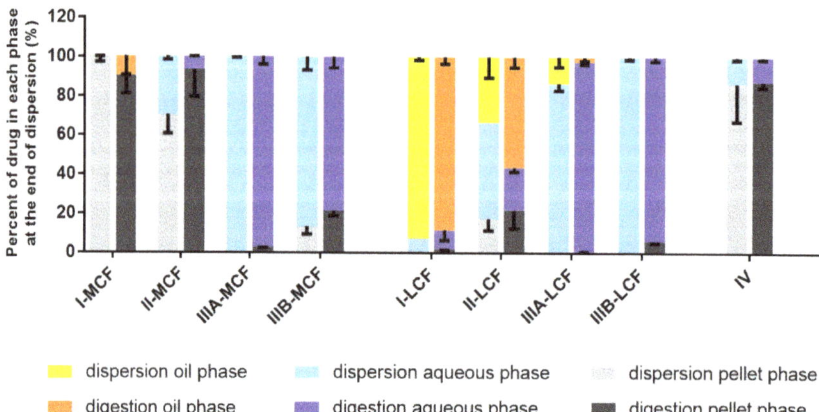

Figure 4. Drug distribution of lumefantrine docusate during in vitro dispersion and digestion tests. Data presents % drug distribution into each phase at the end of dispersion phase and at the end of digestion phase. Data are mean ± SD, $n = 3$.

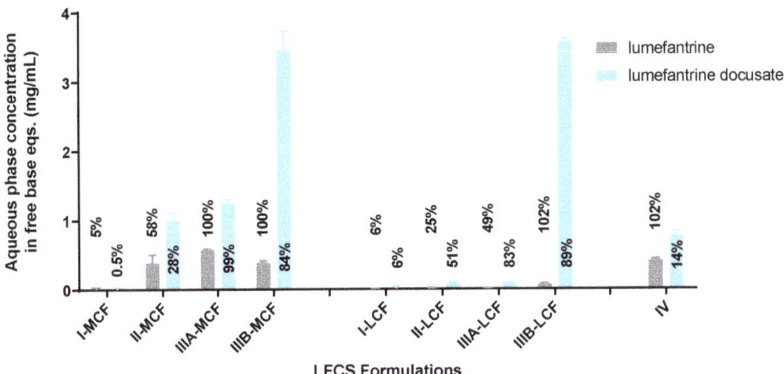

Figure 5. Aqueous phase concentrations of lumefantrine (in free base equivalents) at the end of dispersion phase for LBFs containing lumefantrine or lumefantrine docusate. Percentages are a reflection of the amount of drug present in the aqueous phase compared to total drug loading. Data are mean ± SD, $n = 3$.

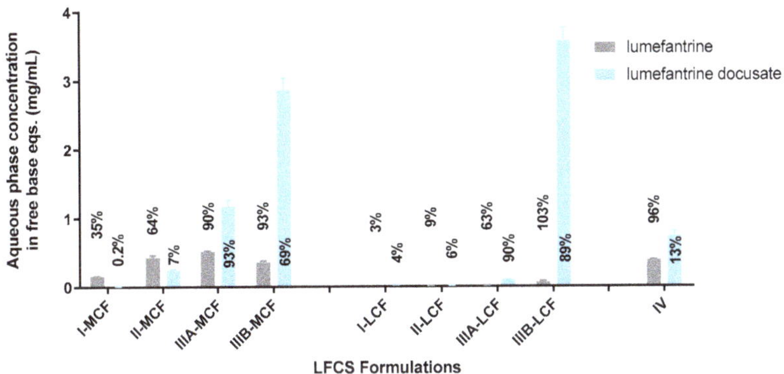

Figure 6. Aqueous phase concentration of lumefantrine (in free base equivalents) at the end of digestion phase for LBFs containing lumefantrine or lumefantrine docusate. Percentages are a reflection of the amount of drug present in the aqueous phase compared to total drug loading. Data are mean ± SD, $n = 3$.

The absolute aqueous phase concentrations (rather than percent distribution) obtained after dispersion and digestion of the different formulations are provided in Figures 5 and 6, respectively. The concentrations reflect the product of the drug loading in the formulation (which in turn is dictated by the solubility data in Figure 2) and the percent solubilized in the aqueous phase. In all cases, the greater solubility of lumefantrine docusate in the different LBF resulted in higher drug concentrations in the aqueous phase (2-to-55-fold higher) than the equivalent lumefantrine free base formulations. The highest aqueous phase concentrations were attained for the medium and long-chain Type IIIB formulations. This reflects the formulations with the highest drug solubility and good ongoing solubilization under digestion conditions. The Type IIIA formulations resulted in good solubilization post digestion, but were limited by drug solubility in the formulation, and the Type IV formulation showed good solubility in the formulation, but more extensive precipitation post digestion.

3.4. *Supersaturation Ratio Calculation for Lumefantrine*

The supersaturation ratio (SR) provides an indication of thermodynamic activity and the likelihood of either drug precipitation, or enhanced absorption from an LBF, where the higher the number, the more supersaturated and therefore the more likely to precipitate during in vitro digestion. The maximum supersaturation ratio (SR^M) provides a measure of the maximum supersaturation pressure that would be generated by digestion of an LBF, assuming no precipitation and is calculated from the following:

$$SR^M = \frac{AP_{max}}{AP_{digest}} \quad (1)$$

where AP_{max} is the maximum possible aqueous phase concentration of drug during the in vitro dispersion and digestion test (i.e., in the absence of any precipitation), and AP_{digest} is the equilibrium solubility of the drug in blank aqueous phase (i.e., where the aqueous phase was obtained from a digestion of blank LBF). Williams et al. have previously proposed that sustainable supersaturation most commonly occurs when $SR^M < 3$, and that exceeding this threshold increases the likelihood of drug precipitation during in vitro dispersion and digestion tests [43]. The concentrations of lumefantrine docusate during in vitro tests for the four formulations are shown in Figure 7. Only the Type II-MC, IIIB-MC, IIIB-LC, and IV formulations were evaluated as these were the four formulations chosen to proceed to the in vivo bioavailability studies. The figures also depict AP_{max}, as well as the drug solubility in the aqueous colloidal phase of the blank LBF (AP_{digest}). In all cases, SR^M was higher than 3, suggesting the potential for precipitation. This was most apparent at early time points for the Type II-MCF formulation where SR^M was ~20 and at later time points post digestion for the Type IIIB-MCF where SR^M was >40. However, unlike previous studies, here the SR^M did not appear to correlate well with drug precipitation (i.e., the presence of drug in the pellet phase). For example, while the SR^M during dispersion was lowest for the Type IIIB-LCF and Type IIIB-MCF, and these two formulations appeared to most robustly resist precipitation, closer analysis of the data shows that for the Type IIIB-LCF, the SR^M at the end of digestion was similar to that of the Type II-MCF and Type IV (where precipitation was significant), but the Type IIIB-LCF was able to maintain drug solubilization throughout the test (89% drug solubilized). Similarly, the SR^M for the Type IIIB-MCF at the end of digestion was the highest, but majority of lumefantrine docusate remained solubilized at 60 min (69% solubilized).

Supersaturation data at 0, 5, 30, and 60 min are shown in Figure 8. The supersaturation ratios in Figure 8 were calculated via the ratio of the concentration of drug measured in the aqueous phase during the in vitro tests compared to the drug solubility in the colloidal aqueous phase at the same time point. This gives an indication of the degree of supersaturation of the solubilized drug concentrations throughout the in vitro test, rather than the maximum degree of supersaturation. The Type II-MCF had a reasonably high SR during dispersion (5.5) and consistent with an SR > 3, the majority of the drug was present in the pellet phase which was likely a mix of phase separation and precipitation. On initiation of digestion, precipitation continued, and the solubilized concentration dropped towards the equilibrium solubility (Figure 7). As such, supersaturation was low for the rest of the test (Figure 8). Similarly, the Type IV formulation dispersion led to an SR^M of 6.3 and significant precipitation on dispersion and digestion. In this case, precipitation to the equilibrium solubility occurred almost immediately and no supersaturation was evident throughout the in vitro test. In contrast for both Type III formulations, solubilization and supersaturation were maintained to varying degrees throughout the in vitro test. For the IIIB LC formulation, SR^M on dispersion was quite low (3.6), and close to the previously described limit for precipitation. Consistent with these previous studies, drug precipitation from this formulation was low and moderate levels of supersaturation were maintained throughout formulation dispersion and digestion. The degree of supersaturation gradually increased throughout the test. For the Type IIIB MC formulation, SR^M on dispersion was slightly higher (5.7) and some precipitation was evident. However, precipitation was still quite low and drug solubilization and supersaturation were maintained. Indeed, supersaturation increased significantly throughout the digestion period due to a fall in equilibrium solubilization capacity. Thus, solubilization and supersaturation were highest for the Type IIIB formulations and amongst these, supersaturation was seemingly higher for the Type IIIB MC formulation.

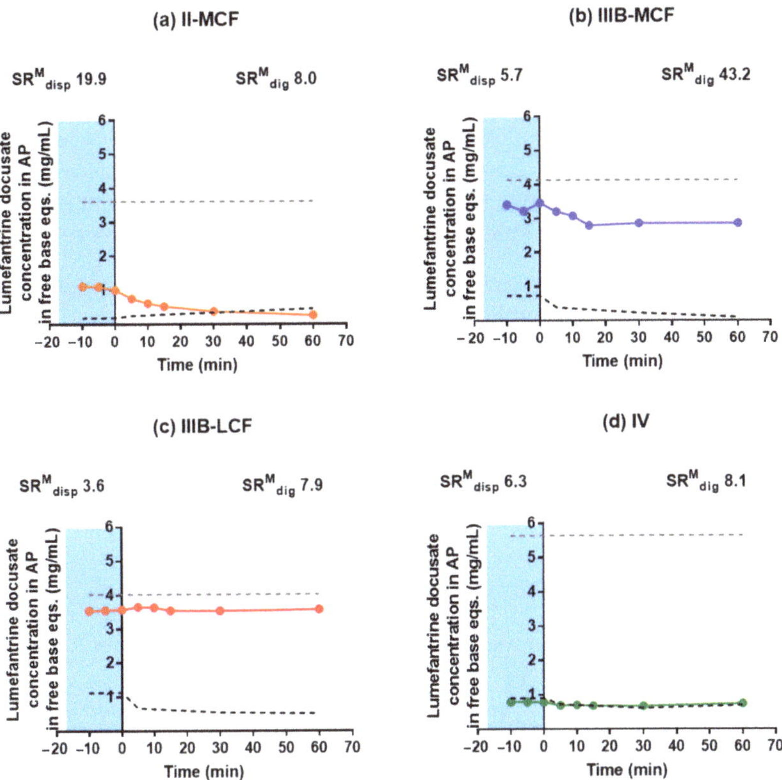

Figure 7. Drug concentrations during in vitro tests (colored circles) of (**a**) Type II-MCF (orange), (**b**) Type IIIB-MCF (blue), (**c**) Type IIIB-LCF (red), and (**d**) Type IV formulations (green). The upper grey line represents the AP_{max}, and the lower dotted black line represents the equilibrium solubility of lumefantrine docusate in the aqueous colloidal phase of blank LBF. The blue shaded section denotes the dispersion phase. Data are mean ± SD, n = 3. The SR^M at time 0 min ($SR^M{}_{dig}$) and 60 min ($SR^M{}_{dig}$) min are displayed for each formulation.

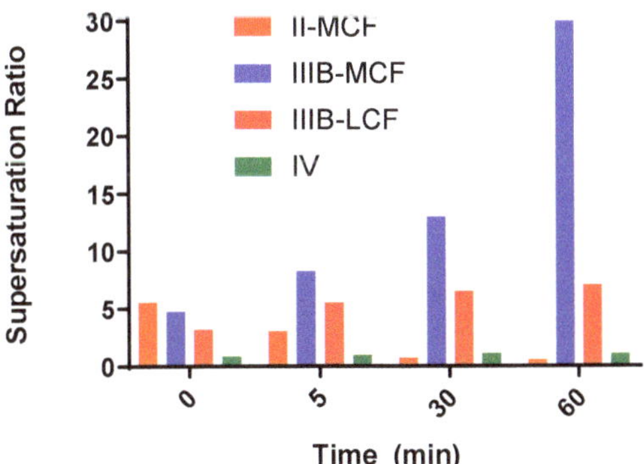

Figure 8. The supersaturation ratios for Type II-MC, Type IIIB-MC, Type IIIB-LC, and Type IV formulations, calculated from the aqueous phase concentrations of the drug during in vitro dispersion and digestion tests at time points of 0, 5, 30, and 60 min divided by the equilibrium solubility of lumefantrine docusate in aqueous colloidal phase of blank LBF at the same times (using data from Figure 7).

3.5. In Vivo Evaluation of Lumefantrine Docusate

After screening the formulations through in vitro dispersion and digestion testing, four lumefantrine docusate formulations were chosen to progress to an in vivo bioavailability study (Type II-MCF, IIIB-MCF, IIIB-LCF, and IV), and the data are shown in Figure 9. The dose for each IL formulation was set at ~80% of the equilibrium solubility value described in Figure 2. The doses of lumefantrine docusate in Type II-MCF, IIIB-MCF, IIIB-LCF, and IV formulations (in free base equivalents) were 50 mg/kg, 65 mg/kg, 65 mg/kg, and 85 mg/kg, respectively. Due to the low lipid solubility of lumefantrine free base, the free base was dosed as a suspension in aqueous and lipid (Type IIIB LCF) formulations at an equivalent dose to the highest IL-LBF (85 mg/kg, attained in the Type IV formulation). The suspension formulations significantly underperformed the equivalent lipid solution formulations made possible by the presence of lumefantrine docusate. Data dose normalized to a fixed nominal dose of 50 mg/kg are shown in the supplementary information, however the major conclusions are unchanged since the two best performing formulations (Type IIIB) were administered at the same dose. The non-normalized data also serve to show the maximal benefit of the IL formulations since this is derived from the ability to both increase dose and harness the advantages of a LBF.

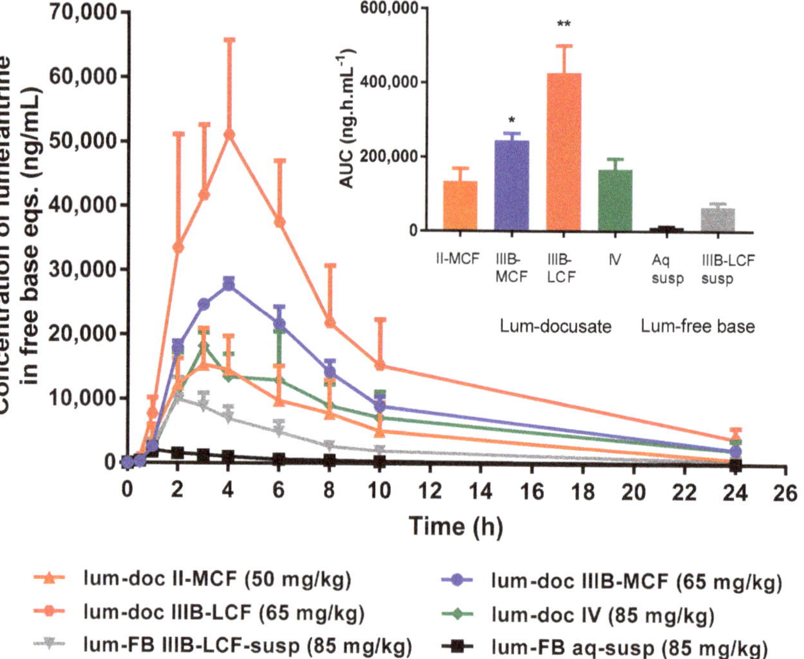

Figure 9. Lumefantrine in vivo bioavailability study. Lumefantrine plasma concentration versus time after oral administration of lumefantrine docusate in Type II-MCF, IIIB-MCF, IIIB-LCF, and IV LBF, as well as lumefantrine free base as an aqueous suspension, and a lipid suspension (in the Type IIIB-LCF). Lumefantrine docusate in Type II-MCF was dosed at 50 mg/kg. Lumefantrine docusate in Type IIIB-MCF and IIIB-LCF were dosed at 65 mg/kg. Lumefantrine docusate in Type IV formulation and the free base suspensions were dosed at 85 mg/kg. Data have been dose normalized to the nominal (i.e., 50–85 mg/kg) dose. Data represented as mean (n = 4) ± SEM. Insert: The total exposure over 24 h, measured as area under the plasma concentration versus time curve from 0 to 24 h. Data are mean ± SD (n = 4). * Formulation was statistically significant ($p < 0.05$) when compared to both suspensions. ** Formulation was statistically significant ($p < 0.05$) from all other formulations.

4. Discussion

Physicochemical Properties of Lumefantrine Compounds. As expected, the melting point of lumefantrine hydrochloride (180–200 °C) was the highest of the lumefantrine salts, followed by lumefantrine free base (133–140 °C), lumefantrine dodecyl sulfate (120–128 °C) and lumefantrine docusate (52–60 °C). The hydrochloride salt, free base, and dodecyl sulfate salt showed birefringence under cross-polarized light, indicating crystallinity, while the docusate-IL did not display birefringence, indicating that the material was amorphous. This property is reflected in the differences in melting point, with the crystalline materials having higher melting points. Higher melting compounds typically have stronger electrostatic forces holding the ions together. In these studies, the molecular bulk of the dodecyl sulfate and docusate counterions was expected to decrease packing efficiency, decrease the strength of intermolecular interactions and therefore decrease the melting point, and this appeared to occur [18]. The greater reduction with the docusate counterion is consistent with a higher molecular weight and enhanced steric bulk of the docusate counterion resulting in a greater disruption of the packing of the crystal lattice than the dodecyl sulfate or the hydrochloride. The difference in melting point between lumefantrine dodecyl sulfate and lumefantrine docusate is similar to the difference between the melting points of the sodium salts of the counterions of ~60 °C (sodium dodecyl sulfate melting point is 205.5 °C, [44,45] sodium docusate melting point is 153–157 °C) [45,46].

Effect of Formulation Type on LBF Solubility of Lumefantrine Compounds. The equilibrium solubility of lumefantrine free base was higher in the more lipophilic formulations (Type I and II formulations), while the HCl salt, dodecyl sulfate salt, and docusate IL all had higher solubilities in the more hydrophilic formulations (Type IIIB and IV formulations). The docusate IL was miscible in the Type IV formulation (i.e., soluble at >1:1 *w*/*w* proportions, the data in Figure 2 reflecting the measured concentration after mixing at a 1:1 ratio). This trend is consistent with previous work, suggesting greater affinity of the drug for the surfactant and co-solvent rich Type III and IV formulations rather than the Type I or Type II lipid rich formulations [35,47]. Thus, even though pairing with highly lipophilic counterions to form an IL might be expected to increase the affinity of lumefantrine for the more lipid rich formulations, the trends in relative solubility across formulation type were similar to that of lumefantrine HCl, and the solubility was higher in the more polar formulations. Surprisingly, lumefantrine dodecyl sulfate only resulted in a solubility advantage compared to lumefantrine free base in the Type IIIB and Type IV formulations, and these increases were only moderate. This may suggest that the dodecyl chain is unable to disrupt packing in the crystal lattice sufficiently to impact solubility significantly. This is consistent with the fact that the reduction in melting point for the dodecyl sulfate was not as pronounced as for the docusate-IL (where solubility was much higher in all formulations). As has been described previously for other drugs, for all lumefantrine salts, drug solubility was higher in the medium-chain formulations when compared to long-chain formulations [48]. This has previously been suggested to reflect the fact that for a fixed mass of lipid, medium-chain lipids contain a greater number of moles of lipid than the long-chain equivalent and drug solubility in lipids appears to be related to the number of ester bonds present in the lipids [48].

Behavior of LBF of Lumefantrine Free Base and Lumefantrine Docusate *in vitro*. The relatively moderate changes to solubility apparent with the dodecyl sulfate salt dictated that further analysis was conducted only with the docusate IL in comparison to lumefantrine free base. Interestingly, the relative solubilization/partition behavior of most of the LBF of lumefantrine and lumefantrine docusate were quite similar after dispersion and digestion, in spite of the much higher drug loading of the IL-containing formulations. For the Type I-MCF, II-MCF and IV formulations, however, differences were evident and a greater proportion of drug was present in the pellet phase for the (higher loaded) lumefantrine docusate formulations. More detailed discussion of the in vitro solubilization trends is provided below.

Type I LBFs are composed of lipids alone and were not fully digested at the end of the in vitro experiment. The lack of surfactant in the formulation also reduced dispersion, resulting in the majority of lumefantrine free base residing in the oil phase of the MCF. In contrast, for the docusate IL, the majority of the drug was present in the pellet phase after digestion of the Type I MCF. This is likely due to phase separation of amorphous lumefantrine IL, rather than precipitation and may reflect the higher loading of drug. Additionally, digestion of medium-chain lipids may yield hydrophobic digestion products which retain the solubilization capacity of the digested formulation, but are denser than the aqueous layer and therefore centrifuge to the bottom of the tube [40]. For both the free base and the docusate IL, the majority of the drug was also present in the oil phase after digestion of the Type I-LCF. In this case, however, this was due to incomplete dispersion and digestion of the long-chain lipids.

The Type II formulations contain a lipophilic surfactant, Tween 85®, which is denser than water. This density difference resulted in a phase separation of the formulation during the dispersion phase for the Type II formulations and recovery of a large proportion of drug in a dense oily pellet phase, particularly for the IL in Type II-MCF. The magnified effect with the IL containing formulation again likely reflects the higher drug loading. Upon digestion, drug distribution across the different phases did not change significantly relative to the dispersion phase, although there was an increased proportion of drug in the pellet phase. As such, drug concentrations in the aqueous phase were low. This increased proportion of drug in the pellet phase is likely a combination of phase separation and precipitation, due to both the denser surfactant and the increased drug loading.

The Type IIIA and IIIB formulations resulted in markedly improved drug solubilization during formulation dispersion and digestion and more than 80% of drug remained solubilized, regardless of the use of lumefantrine or lumefantrine docusate or medium or long-chain lipids. As such, the aqueous phase concentrations obtained were driven by drug solubility in the formulation—which was highest for the Type IIIB formulations, and significantly higher for the IL based formulations relative to the free base.

Type IV formulations are composed entirely of co-solvent and surfactant and resulted in the highest drug solubility in the formulation. The majority of lumefantrine free base remained solubilized at the end of the digestion phase for the Type IV formulation. However, for lumefantrine docusate, there was significantly increased drug precipitation, presumably reflecting the much higher drug loading in the formulation. Type IV formulations containing lumefantrine docusate resulted in substantial precipitation upon dispersion, likely as a result of dilution of co-solvent and surfactant and loss of solvent capacity on dilution. Digestion of the surfactant, Kolliphor® RH 40, may also have caused a further decrease in solubilization capacity. Nonetheless, the much higher drug loading of the docusate-IL in the Type IV LBF resulted in a net effect of much higher aqueous phase drug concentrations when compared to the free base, despite the increased precipitation.

SR^M is an indicator of the propensity of a formulation to precipitate during in vitro dispersion and digestion, where the higher the SR^M, the more likely a formulation is to precipitate. In contrast where the SR^M is below 3, previous results suggest that the formulation is more likely to exhibit sustained supersaturation [41]. This in turn is more likely to generate conditions where absorption is favored in vivo; even relatively brief periods of high supersaturation have been shown to very effectively support absorption for highly permeable drugs [47,49].

The Type II-MCF had the highest SR^M during the dispersion phase, which then decreased upon digestion. The high apparent supersaturation during dispersion is consistent with the majority of the drug being recovered in the pellet phase, which was likely a mix of phase separation and precipitation. The high density of Tween 85® leads to phase separation of surfactant and therefore decreased solubilization capacity. Digestion did not improve the solubilization capacity as the concentration of solubilized drug dropped to the equilibrium solubility of the drug, and therefore the apparent supersaturation was low for the remainder of the in vitro test.

The medium and long chain Type IIIB formulations had a lower SR^M during dispersion, but this increased upon digestion. Both formulations were able to maintain drug solubilization throughout the in vitro test. The degree of apparent supersaturation increased as the in vitro test progressed, with the degree of supersaturation of the Type IIIB-MCF increasing at a greater rate than the LCF counterpart. As medium-chain lipids are more readily digested than the equivalent long-chain lipids [50], the solubilization capacity for medium-chain formulations is lost more quickly than for long-chain formulations, and therefore the degree of supersaturation is greater. The Type IIIB-LCF was able to retain a greater proportion of drug in the solubilized state, consistent with the lower degree of supersaturation compared to the Type IIIB-MCF. In both cases, solubilization was the highest for the Type IIIB formulations. The Type IV formulation had a similar solubilization profile to the Type II-MCF. The SR^M upon dispersion was 6.3, which then increased to 8.1 at the end of digestion. Unlike the Type IIIB formulations, however, the majority of the drug precipitated out during dispersion of the Type IV formulation and concentrations remained at equilibrium solubility throughout the in vitro test. As such, no supersaturation was observed.

Although supersaturation is a driver of precipitation *in vitro*, increases in supersaturation in vivo result in increases in thermodynamic activity and may therefore result in increases in absorption. As such, whether supersaturation results in an increase or a decrease in absorption is typically a trade-off between the drivers of precipitation and absorption. The in vitro data suggest that the Type III formulations, where both solubilization and supersaturation were maintained, were most likely to promote absorption in vivo. This suggestion was subsequently probed via the conduct of an in vivo bioavailability study.

Effect of Formulation Type on in vivo Bioavailability of Lumefantrine Docusate. Oral administration of all four lumefantrine docusate IL-containing solution LBFs resulted in higher systemic exposure than that obtained after oral administration of aqueous or lipid suspensions of lumefantrine free base (Figure 9). Statistical analysis by one-way ANOVA suggest significantly higher exposure after administration of the docusate IL containing Type IIIB-MCF compared to both free base suspensions, while the Type IIIB-LCF exhibited significantly higher exposure than all other formulations tested. Administration of the lumefantrine docusate Type IIIB-LCF resulted in the highest plasma concentrations, followed by the Type IIIB-MCF. The Type IV and II-MCF formulations had similar and intermediate exposure profiles, and these were both higher than the lipid suspension and aqueous suspension of the free base. Figure 9 shows these trends and illustrates that the IL containing formulations that are able to dissolve higher quantities of drug and therefore facilitate the administration of higher doses also result in higher exposure. The data also show that for lumefantrine, lipid suspension formulations were unable to provide the same benefits to exposure apparent with the IL containing lipid solution formulations. The profiles were also normalized to the same dose, to remove the effects of differing dose and to instead look at the intrinsic absorption promoting ability of the formulations at a fixed dose. Surprisingly, however, this did not markedly change the conclusions, except to further relegate the Type IV formulation where dose was high, but precipitation was also very high. Thus, although the Type IV formulation allows for the highest drug loading, the high dose is not enough to overcome the precipitation that occurs upon administration (and indeed may stimulate it). The exposure obtained after administration of the Type II-MCF was also relatively poor, reflecting both low drug loading and low absorption enhancement. In contrast, and broadly consistent with the in vitro data, both Type IIIB formulations resulted in significantly higher lumefantrine exposure after oral administration. The use of the IL therefore allowed for both higher doses and enhanced exposure (up to 50-fold) when compared to the free base.

Correlating *in vitro-in vivo* performance is complex as there are many factors which affect performance. For example, the lack of an absorption sink in vitro usually results in overestimation of precipitation [38,50]. Choice of in vivo model (i.e., rat, pig, dog, etc.) provides an additional layer of complexity as each model has physiological differences that make prediction of the eventual translation to humans difficult [51]. Nevertheless, there have been examples of lipid formulations (including those that contain IL) that do show a strong *in vitro/in vivo* correlation. For example, Sahbaz et al. have reported that combination of cinnarizine decyl sulfate with a SEDDS formulation resulted in ~2-fold higher exposure in rats when compared to an equivalent dose of cinnarizine free base as a suspension in the same SEDDS formulation. Similarly, itraconazole docusate in a SEDDS formulation resulted in ~20-fold higher exposure in rats than an itraconazole suspension of equal dose, and exhibited greater exposure than the commercial formulation Sporanox®. For both cinnarizine decyl sulfate and itraconazole docusate, the enhanced performance of the IL in vivo was consistent with improved in vitro solubilization on formulation dispersion and digestion when compared to the corresponding free base suspensions [32].

Williams et al. have also reported that lipophilic salt forms of small molecule kinase inhibitors exhibit higher solubility in lipidic excipients when compared to the free base or commercial salt form. In this example, increased solubility resulted in increased drug loading of the small molecule kinase inhibitors in LBF. Isolation of erlotinib and cabozantinib as lipophilic salts (erlotinib docusate and cabozantinib docusate) also led to increased aqueous phase concentrations in vitro at gastric pH when compared to an aqueous suspension of the corresponding hydrochloride salt. Under intestinal conditions where pH increased and digestion was stimulated, solubilized drug concentration dropped significantly, however the IL formulations were still able to provide solubilization advantage when compared to the free base. In vivo, LBF containing erlotinib docusate resulted in lower variability and better dose linearity when compared to an aqueous suspension of the (commercial) hydrochloride salt. Cabozantinib docusate also resulted in improvements in exposure when formulated in an LBF and bioavailability was enhanced 1.5- and 1.8-fold after administration of a medium-chain SEDDS and long-chain SEDDS compared to an aqueous suspension of cabozantinib free base, respectively [52].

To more carefully analyze the relationship between in vitro and in vivo endpoints, Figure 10 shows the correlation between the degree of drug solubilization during in vitro dispersion and digestion (expressed as the AUC of the drug concentration in the digest aqueous phase over time) and in vivo exposure (expressed as the AUC of the plasma drug concentration versus time profile after oral administration). The strength of the correlation suggests that increases in drug solubilization translate to increases in drug absorption and exposure. The drug dose/load in the formulations in both the in vitro and in vivo tests varied across formulations, but was the same in each test (i.e., in vitro and in vivo) for each formulation. According to the Pearson coefficient (2-tailed test), the strength of the association between the in vitro performance and in vivo performance is high ($r^2 = 0.8695$), and the correlation is significant ($p < 0.01$).

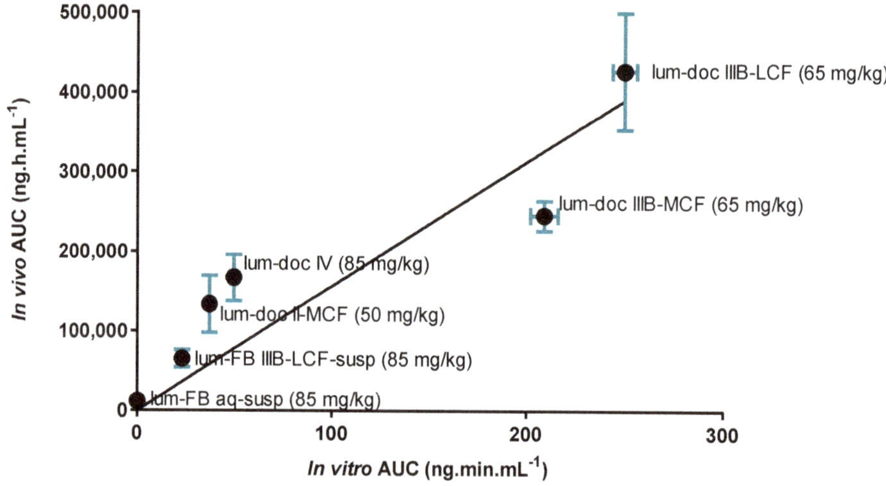

Figure 10. In vitro/in vivo correlation plot displaying the AUC for the aqueous solubilization during in vitro dispersion and digestion experiments, and the AUC for in vivo exposure. Data are expressed as mean ± SD for in vitro AUC ($n = 3$) and mean ± SEM for in vivo AUC ($n = 4$).

For poorly water-soluble drugs, increases in drug solubilization typically promote drug absorption as shown in Figure 10. In addition to drug solubilization, previous studies have also shown that differences in the degree of drug supersaturation (rather than solubilization) may be important drivers of absorption [45,47]. In the current studies, however, this appears not to be the case as there is no correlation between supersaturation and drug exposure (Supplementary Figure S20). The data therefore suggest that for lumefantrine, solubilization behavior appears to be a better indicator of in vivo performance (Figure 10) than supersaturation. This trend contrasts with recent studies of similar lipid formulations of fenofibrate, where it was reported that supersaturation was a more significant driver of in vivo exposure [47,49]. However, Crum et al. found that for higher doses of fenofibrate, where longer absorption periods were required, ongoing solubilization was the more significant driver of absorption [47]. As the drug loading for lumefantrine docusate is relatively high, longer absorption times may be required and therefore ongoing solubilization may drive absorption more effectively than supersaturation. The difference in the driving force for absorption may also reflect differences in intrinsic permeability, especially at lower dose. Thus, for drugs which are absorbed quickly, and where permeability is high such as fenofibrate, relatively brief periods of supersaturation may be very effective drivers of in vivo absorption. In contrast, where absorption is slower, permeability is lower and dose is higher, for example with lumefantrine, ongoing solubilization may be more important.

5. Conclusions

The use of API-ILs in conjunction with lipid-based formulations has been examined as a means to enhance the oral exposure of lumefantrine. The data suggest significant benefits in solubility in lipid-based formulations and oral exposure are possible using this approach. Isolation as lumefantrine docusate resulted in consistent increases in lipid solubility when compared to lumefantrine free base across a range of lipid-based formulations. The benefits in solubility were subsequently shown to persist and provide for performance advantages during in vitro dispersion and digestion testing, and ultimately, exposure in vivo. The data suggest that using a large, bulky, lipophilic counterion (such as docusate) can both improve the lipid solubility of the parent compound and enhance drug solubilization during formulation processing under simulated GI conditions. For the first time for LBF of ILs, a range of different formulations were explored and the Type IIIB LBFs resulted in the most effective solubilization and supersaturation in vitro and almost completely resisted precipitation. In vivo, these two formulations also significantly out-performed Type II and Type IV formulations of lumefantrine. Within the Type IIIB formulations, the LCF based formulation appeared to support more effective absorption of lumefantrine when compared to the Type IIIB MCF. This was consistent with slightly better drug solubilization for the Type IIIB LCF formulation, but was in contrast to the lower levels of supersaturation when compared to the Type IIIB MC formulation. The outperformance of the LC lipid containing formulation may also reflect improved support for lymphatic transport since the close structural analogue, halofantrine, [53,54] has been shown previously to be highly lymphatically transported, and LC lipids more effectively support lymph transport than medium chain lipids [55]. However, increases in lymphatic transport are thought to increase bioavailability primarily via changes in first pass metabolism and the potential importance of high first pass metabolism relative to low water solubility in driving the low oral bioavailability of lumefantrine is not known. Nonetheless, in all cases, the use of the docusate IL was able to significantly enhance drug loading in lipid formulations, and promote solubilization. In the case of the Type IIIB LC formulation, this solubilization advantage resulted in very large increases in exposure after oral administration when compared to aqueous (~50 fold) or lipid (~10 fold) suspensions of the free base. The data lend further support to the potential utility of ionic liquid drug forms as a means to enhance drug exposure [1,33,56].

Supplementary Materials: The following are available online at http://www.mdpi.com/1999-4923/12/1/17/s1. ^{1}H, ^{13}C NMR and HR-MS Spectra/Data. Figure S1: In vitro dispersion and digestion of lumefantrine free base in Type I-MCF. The concentration of lumefantrine free base in the aqueous phase of the medium-chain digest as a function of time (left), and the proportion of lumefantrine free base in the pellet, aqueous, and oil phases as a function of time (right). Data are $n = 3$, mean ± SD. * Figure S2: In vitro dispersion and digestion of lumefantrine free base in Type I-LCF. Figure S3: In vitro dispersion and digestion of lumefantrine free base in Type II-MCF. Figure S4: In vitro dispersion and digestion of lumefantrine free base in Type II-LCF. Figure S5: In vitro dispersion and digestion of lumefantrine free base in Type IIIA-MCF. Figure S6: In vitro dispersion and digestion of lumefantrine free base in Type IIIA-LCF. Figure S7: In vitro dispersion and digestion of lumefantrine free base in Type IIIB-MCF. Figure S8: In vitro dispersion and digestion of lumefantrine free base in Type IIIB-LCF. Figure S9: In vitro dispersion and digestion of lumefantrine free base in Type IV. Figure S10: In vitro dispersion and digestion of lumefantrine docusate in Type I-MCF. Figure S11: In vitro dispersion and digestion of lumefantrine docusate in Type I-LCF. Figure S12: In vitro dispersion and digestion of lumefantrine docusate in Type II-MCF. Figure S13: In vitro dispersion and digestion of lumefantrine docusate in Type II-LCF. Figure S14: In vitro dispersion and digestion of lumefantrine docusate in Type IIIA-MCF. Figure S15: In vitro dispersion and digestion of lumefantrine docusate in Type IIIA-LCF. Figure S16: In vitro dispersion and digestion of lumefantrine docusate in Type IIIB-MCF. Figure S17: In vitro dispersion and digestion of lumefantrine docusate in Type IIIB-LCF. Figure S18: In vitro dispersion and digestion of lumefantrine docusate in Type IV. Figure S19: Dose normalized lumefantrine plasma concentration versus time data after oral administration of lumefantrine docusate in Type II-MCF, IIIB-MCF, IIIB-LCF, and IV LBF, as well as lumefantrine free base as an aqueous suspension, and a lipid suspension (in the Type IIIB-LCF). All data have been dose normalized to 50 mg/kg (reflecting the lowest dose administered (Type II-MCF)). Data represented as mean (n = 4) ± SEM. Insert: Total lumefantrine exposure over 24 h. Data represented as mean (n = 4) ± SEM. * Exposure was statistically higher ($p < 0.05$) than both suspension formulations. ** Exposure was statistically higher than all other formulations. Figure S20: Apparent supersaturation ratio/in vivo correlation plot, displaying the AUC of the apparent supersaturation ratio across the in vitro experiment, and the AUC for in vivo exposure. Data are expressed as mean ± SEM for in vivo AUC (n = 4). * The details for Figures S2–S18 are the same, but the extended titles have been left out for the sake of brevity.

Author Contributions: Conceptualization, C.J.H.P., P.J.S., H.D.W. and H.B.; methodology, E.T., T.-H.N., and L.F.; investigation, E.T.; writing—original draft preparation, E.T.; writing—review and editing, E.T., T.-H.N., L.F., P.J.S., and C.J.H.P.; Supervision, T.-H.N., L.F., P.J.S., and C.J.H.P. All authors have read and agreed to the published version of the manuscript.

Funding: This research was in part funded by Lonza. Author E.T. received scholarship support from the Australian Government and Monash University Faculty of Pharmacy and Pharmaceutical Sciences.

Acknowledgments: The authors wish to thank Yusuke Tanaka for assistance with the in vivo studies and Jason Dang for acquiring all HRMS data.

Conflicts of Interest: This work describes intellectual property in the use of ionic liquids/lipophilic salts in drug delivery that has been assigned to Lonza from Monash University. H.D.W., H.B. and L.F. are all employees of Lonza and participated in the conduct or write up of this study.

References

1. Stoimenovski, J.; MacFarlane, D.R.; Bica, K.; Rogers, R.D. Crystalline vs. Ionic liquid salt forms of active pharmaceutical ingredients: A position paper. *Pharm. Res.* **2010**, *27*, 521–526. [CrossRef] [PubMed]
2. Leeson, P.D. Molecular inflation, attrition and the rule of five. *Adv. Drug Deliv. Rev.* **2016**, *101*, 22–33. [CrossRef] [PubMed]
3. Bergstrom, C.A.S.; Porter, C.J.H. Computational prediction of formulation strategies for beyond-rule-of-five compounds. *Adv. Drug Deliv. Rev.* **2016**, *101*, 6–21. [CrossRef] [PubMed]
4. Tsume, Y.; Mudie, D.M.; Langguth, P.; Amidon, G.E.; Amidon, G.L. The biopharmaceutics classification system: Subclasses for in vivo predictive dissolution (IPD) methodology and IVIVC. *Eur. J. Pharm. Sci.* **2014**, *57*, 152–163. [CrossRef] [PubMed]
5. Pouton, C.W. Formulation of poorly water-soluble drugs for oral administration: Physicochemical and physiological issues and the lipid formulation classification system. *Eur. J. Pharm. Sci.* **2006**, *29*, 278–287. [CrossRef] [PubMed]
6. Williams, H.D.; Sahbaz, Y.; Ford, L.; Nguyen, T.-H.; Scammells, P.J.; Porter, C.J.H. Ionic liquids provide unique opportunities for oral drug delivery: Structure optimization and in vivo evidence of utility. *Chem. Commun.* **2014**, *50*, 1688–1690. [CrossRef] [PubMed]
7. Williams, H.D.; Trevaskis, N.L.; Charman, S.A.; Shanker, R.M.; Charman, W.N.; Pouton, C.W.; Porter, C.J.H. Strategies to address low drug solubility in discovery and development. *Pharmacol. Rev.* **2013**, *65*, 315–499.
8. Holm, R. Bridging the gaps between academic research and industrial product developments of lipid based formulations. *Adv. Drug Deliv. Rev.* **2019**, *142*, 118–127. [CrossRef]
9. Davis, M.; Walker, G. Recent strategies in spray drying for the enhanced bioavailability of poorly water soluble drugs. *J. Control. Release* **2018**, *269*, 110–127. [CrossRef]
10. Makanga, M.; Krudsood, S. The clinical efficacy of artemether/lumefantrine (coartem®). *Malar. J.* **2009**, *8*, S5. [CrossRef]

11. Patel, K.; Sarma, V.; Vavia, P. Design and evaluation of lumefantrine – oleic acid self nanoemulsifying ionic complex for enhanced dissolution. *DARU J. Pharm. Sci.* **2013**, *21*, 27–37. [CrossRef] [PubMed]
12. Zhao, Y.L.; Liu, X.M.; Lu, X.M.; Zhang, S.J.; Wang, J.J.; Wang, H.; Gurau, G.; Rogers, R.D.; Su, L.; Li, H.N. The behavior of ionic liquids under high pressure: A molecular dynamics simulation. *J. Phys. Chem. B* **2012**, *116*, 10876–10884. [CrossRef] [PubMed]
13. Bica, K.; Rogers, R.D. Confused ionic liquid ions-a "liquification" and dosage strategy for pharmaceutically active salts. *Chem. Commun.* **2010**, *46*, 1215–1217. [CrossRef] [PubMed]
14. Bica, K.; Rijksen, C.; Nieuwenhuyzena, M.; Rogers, R.D. In search of pure liquid salt forms of aspirin: Ionic liquid approaches with acetylsalicylic acid and salicylic acid. *Phys. Chem. Chem. Phys.* **2010**, *12*, 2011–2017. [CrossRef]
15. Dobler, D.; Schmidts, T.; Klingenhofer, I.; Runkel, F. Ionic liquids as ingredients in topical drug delivery systems. *Int. J. Pharm.* **2013**, *441*, 620–627. [CrossRef]
16. Qamar, S.; Brown, P.; Ferguson, S.; Khan, R.A.; Ismail, B.; Khan, A.R.; Sayed, M.; Khan, A.M. The interaction of a model active pharmaceutical with cationic surfactant and the subsequent design of drug based ionic liquid surfactants. *J. Colloid Interface Sci.* **2016**, *481*, 117–124. [CrossRef]
17. Arsalan, M.; Richard, A.O.B.; Samuel, M.M.; Kaila, M.M.; Niloufar, M.; Kevin, N.W.; James, H.D. Lipid-inspired ionic liquids containing long-chain appendages: Novel class of biomaterials with attractive properties and applications. In *Ionic Liquids: Science and Applications*; American Chemical Society: Washington, DC, USA, 2012; Volume 1117, pp. 199–216.
18. Katritzky, A.R.; Jain, R.; Lomaka, A.; Petrukhin, R.; Maran, U.; Karelson, M. Perspective on the relationship between melting points and chemical structure. *Cryst. Growth Des.* **2001**, *1*, 261–265. [CrossRef]
19. Shamshina, J.L.; Barber, P.S.; Rogers, R.D. Ionic liquids in drug delivery. *Expert Opin. Drug Deliv.* **2013**, *10*, 1367–1381. [CrossRef]
20. Balk, A.; Wiest, J.; Widmer, T.; Galli, B.; Holzgrabe, U.; Meinel, L. Transformation of acidic poorly water soluble drugs into ionic liquids. *Eur. J. Pharm. Biopharm.* **2015**, *94*, 73–82. [CrossRef]
21. Balk, A.; Widmer, T.; Wiest, J.; Bruhn, H.; Rybak, J.-C.; Matthes, P.; Müller-Buschbaum, K.; Sakalis, A.; Lühmann, T.; Berghausen, J.; et al. Ionic liquid versus prodrug strategy to address formulation challenges. *Pharm. Res.* **2015**, *32*, 2154–2167. [CrossRef]
22. Shadid, M.; Gurau, G.; Shamshina, J.L.; Chuang, B.C.; Hailu, S.; Guan, E.; Chowdhury, S.K.; Wu, J.T.; Rizvi, S.A.A.; Griffin, R.J.; et al. Sulfasalazine in ionic liquid form with improved solubility and exposure. *MedChemComm* **2015**, *6*, 1837–1841. [CrossRef]
23. Phan, T.N.Q.; Shahzadi, I.; Bernkop-Schnurch, A. Hydrophobic ion-pairs and lipid-based nanocarrier systems: The perfect match for delivery of BCS class 3 drugs. *J. Control. Release* **2019**, *304*, 146–155. [CrossRef] [PubMed]
24. Nanjwade, B.K.; Patel, D.J.; Udhani, R.A.; Manvi, F.V. Functions of lipids for enhancement of oral bioavailability of poorly water-soluble drugs. *Sci. Pharm.* **2011**, *79*, 705–727. [CrossRef] [PubMed]
25. Hauss, D.J. Oral lipid-based formulations. *Adv. Drug Deliv. Rev.* **2007**, *59*, 667–676. [CrossRef] [PubMed]
26. Persson, L.C.; Porter, C.J.; Charman, W.N.; Bergstrom, C.A. Computational prediction of drug solubility in lipid based formulation excipients. *Pharm. Res.* **2013**, *30*, 3225–3237. [CrossRef] [PubMed]
27. Mu, H.; Holm, R.; Müllertz, A. Lipid-based formulations for oral administration of poorly water-soluble drugs. *Int. J. Pharm.* **2013**, *453*, 215–224. [CrossRef] [PubMed]
28. Porter, C.J.H.; Trevaskis, N.L.; Charman, W.N. Lipids and lipid-based formulations: Optimizing the oral delivery of lipophilic drugs. *Nat. Rev. Drug Discov.* **2007**, *6*, 231–248. [CrossRef]
29. Rahman, M.A.; Harwansh, R.; Mirza, M.A.; Hussain, S.; Hussain, A. Oral lipid based drug delivery system (lbdds): Formulation, characterization and application: A review. *Curr. Drug Deliv.* **2011**, *8*, 330–345. [CrossRef]
30. Tiwari, R.; Tiwari, G.; Rai, A. Self-emulsifying drug delivery system: An approach to enhance solubility. *Syst. Rev. Pharm.* **2010**, *1*, 133–140. [CrossRef]
31. Singh, A.; Worku, Z.A.; Van den Mooter, G. Oral formulation strategies to improve solubility of poorly water-soluble drugs. *Expert Opin. Drug Deliv.* **2011**, *8*, 1361–1378. [CrossRef]
32. Sahbaz, Y.; Williams, H.D.; Nguyen, T.H.; Saunders, J.; Ford, L.; Charman, S.A.; Scammells, P.J.; Porter, C.J. Transformation of poorly water-soluble drugs into lipophilic ionic liquids enhances oral drug exposure from lipid based formulations. *Mol. Pharm.* **2015**, *12*, 1980–1991. [CrossRef] [PubMed]

33. Williams, H.D.; Ford, L.; Igonin, A.; Shan, Z.; Botti, P.; Morgen, M.M.; Hu, G.; Pouton, C.W.; Scammells, P.J.; Porter, C.J.H.; et al. Unlocking the full potential of lipid-based formulations using lipophilic salt/ionic liquid forms. *Adv. Drug Deliv. Rev.* **2019**, *142*, 75–90. [CrossRef] [PubMed]
34. Morgen, M.; Saxena, A.; Chen, X.Q.; Miller, W.; Nkansah, R.; Goodwin, A.; Cape, J.; Haskell, R.; Su, C.; Gudmundsson, O.; et al. Lipophilic salts of poorly soluble compounds to enable high-dose lipidic SEDDS formulations in drug discovery. *Eur. J. Pharm. Biopharm.* **2017**, *117*, 212–223. [CrossRef] [PubMed]
35. Williams, H.D.; Sassene, P.; Kleberg, K.; Bakala-N'Goma, J.-C.; Calderone, M.; Jannin, V.; Igonin, A.; Partheil, A.; Marchaud, D.; Jule, E.; et al. Toward the establishment of standardized in vitro tests for lipid-based formulations, part 1: Method parameterization and comparison of in vitro digestion profiles across a range of representative formulations. *J. Pharm. Sci.* **2012**, *101*, 3360–3380. [CrossRef] [PubMed]
36. Kohli, K.; Chopra, S.; Dhar, D.; Arora, S.; Khar, R.K. Self-emulsifying drug delivery systems: An approach to enhance oral bioavailability. *Drug Discov. Today* **2010**, *15*, 958–965. [CrossRef]
37. Williams, H.D.; Sassene, P.; Kleberg, K.; Calderone, M.; Igonin, A.; Jule, E.; Vertommen, J.; Blundell, R.; Benameur, H.; Müllertz, A.; et al. Toward the establishment of standardized in vitro tests for lipid-based formulations, part 4: Proposing a new lipid formulation performance classification system. *J. Pharm. Sci.* **2014**, *103*, 2441–2455. [CrossRef]
38. Crum, M.F.; Trevaskis, N.L.; Williams, H.D.; Pouton, C.W.; Porter, C.J.H. A new in vitro lipid digestion—In vivo absorption model to evaluate the mechanisms of drug absorption from lipid-based formulations. *Pharm. Res.* **2016**, *33*, 970–982. [CrossRef]
39. Anby, M.U.; Williams, H.D.; McIntosh, M.; Benameur, H.; Edwards, G.A.; Pouton, C.W.; Porter, C.J.H. Lipid digestion as a trigger for supersaturation: Evaluation of the impact of supersaturation stabilization on the in vitro and in vivo performance of self-emulsifying drug delivery systems. *Mol. Pharm.* **2012**, *9*, 2063–2079. [CrossRef]
40. Lee, K.W.Y.; Porter, C.J.H.; Boyd, B.J. The effect of administered dose of lipid-based formulations on the in vitro and in vivo performance of cinnarizine as a model poorly water-soluble drug. *J. Pharm. Sci.* **2012**, *102*, 565–578. [CrossRef]
41. Govender, K.; Gibhard, L.; Du Plessis, L.; Wiesner, L. Development and validation of a lc–ms/ms method for the quantitation of lumefantrine in mouse whole blood and plasma. *J. Chromatogr. B* **2015**, *985*, 6–13. [CrossRef]
42. Williams, H.D.; Anby, M.U.; Sassene, P.; Kleberg, K.; Bakala-N'Goma, J.-C.; Calderone, M.; Jannin, V.; Igonin, A.; Partheil, A.; Marchaud, D.; et al. Toward the establishment of standardized in vitro tests for lipid-based formulations. 2. The effect of bile salt concentration and drug loading on the performance of type I, II, IIIA, IIIB, and IV formulations during in vitro digestion. *Mol. Pharm.* **2012**, *9*, 3286–3300. [CrossRef] [PubMed]
43. Williams, H.D.; Trevaskis, N.L.; Yeap, Y.Y.; Anby, M.U.; Pouton, C.W.; Porter, C.J. Lipid-based formulations and drug supersaturation: Harnessing the unique benefits of the lipid digestion/absorption pathway. *Pharm. Res.* **2013**, *30*, 2976–2992. [CrossRef] [PubMed]
44. Sodium Dodecyl Sulfate. Drugbank. Available online: https://www.drugbank.ca/drugs/DB00815 (accessed on 18 June 2019).
45. Wishart, D.S.; Knox, C.; Guo, A.C.; Shrivastava, S.; Hassanali, M.; Stothard, P.; Chang, Z.; Woolsey, J. Drugbank: A comprehensive resource for in silico drug discovery and exploration. *Nucleic Acids Res.* **2006**, *34*, D668–D672. [CrossRef] [PubMed]
46. Sodium Docusate. Drugbank. Available online: https://www.drugbank.ca/drugs/DB11089 (accessed on 18 June 2019).
47. Crum, M.F.; Trevaskis, N.L.; Pouton, C.W.; Porter, C.J.H. Transient supersaturation supports drug absorption from lipid-based formulations for short periods of time, but ongoing solubilization is required for longer absorption periods. *Mol. Pharm.* **2017**, *14*, 394–405. [CrossRef]
48. Kaukonen, A.; Boyd, B.; Porter, C.H.; Charman, W. Drug solubilization behavior during in vitro digestion of simple triglyceride lipid solution formulations. *Pharm. Res.* **2004**, *21*, 245–253. [CrossRef]
49. Suys, E.J.A.; Chalmers, D.K.; Pouton, C.W.; Porter, C.J.H. Polymeric precipitation inhibitors promote fenofibrate supersaturation and enhance drug absorption from a type IV lipid-based formulation. *Mol. Pharm.* **2018**, *15*, 2355–2371. [CrossRef]

50. Feeney, O.M.; Crum, M.F.; McEvoy, C.L.; Trevaskis, N.L.; Williams, H.D.; Pouton, C.W.; Charman, W.N.; Bergstrom, C.A.S.; Porter, C.J.H. 50 years of oral lipid-based formulations: Provenance, progress and future perspectives. *Adv. Drug Deliv. Rev.* **2016**, *101*, 167–194. [CrossRef]
51. O'Driscoll, C.M.; Griffin, B.T. Biopharmaceutical challenges associated with drugs with low aqueous solubility—The potential impact of lipid-based formulations. *Adv. Drug Deliv. Rev.* **2008**, *60*, 617–624. [CrossRef]
52. Williams, H.D.; Ford, L.; Han, S.; Tangso, K.J.; Lim, S.F.; Shakleford, D.M.; Vodak, D.T.; Benameur, H.; Pouton, C.W.; Scammells, P.J.; et al. Enhancing the oral absorption of kinase inhibitors using lipophilic salts and lipid based formulations. *Mol. Pharm.* **2018**, *15*, 5678–5696. [CrossRef]
53. Khoo, S.-M.; Shackleford, D.M.; Porter, C.J.H.; Edwards, G.A.; Charman, W.N. Intestinal lymphatic transport of halofantrine occurs after oral administration of a unit-dose lipid-based formulation to fasted dogs. *Pharm. Res.* **2003**, *20*, 1460–1465. [CrossRef]
54. Porter, C.J.H.; Charman, S.A.; Charman, W.N. Lymphatic transport of halofantrine in the triple-cannulated anesthetized rat model: Effect of lipid vehicle dispersion. *J. Pharm. Sci.* **1996**, *85*, 351–356. [CrossRef] [PubMed]
55. Caliph, S.M.; Cao, E.; Bulitta, J.B.; Hu, L.; Han, S.; Porter, C.J.H.; Trevaskis, N.L. The impact of lymphatic transport on the systemic disposition of lipophilic drugs. *J. Pharm. Sci.* **2013**, *102*, 2395–2408. [CrossRef] [PubMed]
56. Huang, W.; Wu, X.; Qi, J.; Zhu, Q.; Wu, W.; Lu, Y.; Chen, Z. Ionic liquids: Green and tailor-made solvents in drug delivery. *Drug Discov. Today* **2019**. [CrossRef] [PubMed]

Article

Ionic Liquid-In-Oil Microemulsions Prepared with Biocompatible Choline Carboxylic Acids for Improving the Transdermal Delivery of a Sparingly Soluble Drug

Md. Rafiqul Islam [1,2], Md. Raihan Chowdhury [1], Rie Wakabayashi [1,3], Noriho Kamiya [1,3,4], Muhammad Moniruzzaman [5] and Masahiro Goto [1,3,4,*]

1 Department of Applied Chemistry, Graduate School of Engineering, Kyushu University, 744 Motooka, Nishi-ku, Fukuoka 819-0395, Japan; islam.md.334@s.kyushu-u.ac.jp (M.R.I.); chowdhury.md.raihan.588@m.kyushu-u.ac.jp (M.R.C.); rie_wakaba@mail.cstm.kyushu-u.ac.jp (R.W.); kamiya.noriho.367@m.kyushu-u.ac.jp (N.K.)

2 Department of Applied Chemistry and Chemical Engineering, Noakhali Science and Technology University, Noakhali 3814, Bangladesh

3 Advanced Transdermal Drug Delivery System Center, Kyushu University, 744 Motooka, Nishi-ku, Fukuoka 819-0395, Japan

4 Division of Biotechnology, Center for Future Chemistry, Kyushu University, 744 Motooka, Nishi-ku, Fukuoka 819-0395, Japan

5 Chemical Engineering Department, Universiti Teknologi PETRONAS, Seri Iskandar 32610, Perak, Malaysia; m.moniruzzaman@utp.edu.my

* Correspondence: m-goto@mail.cstm.kyushu-u.ac.jp; Tel.: +81-92-802-2806; Fax: +81-92-802-2810

Received: 31 March 2020; Accepted: 22 April 2020; Published: 24 April 2020

Abstract: The transdermal delivery of sparingly soluble drugs is challenging due to of the need for a drug carrier. In the past few decades, ionic liquid (IL)-in-oil microemulsions (IL/O MEs) have been developed as potential carriers. By focusing on biocompatibility, we report on an IL/O ME that is designed to enhance the solubility and transdermal delivery of the sparingly soluble drug, acyclovir. The prepared MEs were composed of a hydrophilic IL (choline formate, choline lactate, or choline propionate) as the non-aqueous polar phase and a surface-active IL (choline oleate) as the surfactant in combination with sorbitan laurate in a continuous oil phase. The selected ILs were all biologically active ions. Optimized pseudo ternary phase diagrams indicated the MEs formed thermodynamically stable, spherically shaped, and nano sized (<100 nm) droplets. An in vitro drug permeation study, using pig skin, showed the significantly enhanced permeation of acyclovir using the ME. A Fourier transform infrared spectroscopy study showed a reduction of the skin barrier function with the ME. Finally, a skin irritation study showed a high cell survival rate (>90%) with the ME compared with Dulbecco's phosphate-buffered saline, indicates the biocompatibility of the ME. Therefore, we conclude that IL/O ME may be a promising nano-carrier for the transdermal delivery of sparingly soluble drugs.

Keywords: biocompatible; ionic liquid; transdermal drug delivery system; microemulsion

1. Introduction

The transdermal drug delivery system (TDDS), a safe, non-invasive, easy, and effective drug delivery system, has attracted much attention in recent research because of its numerous prospective advantages, including improved patient compliance, avoidance of the first-pass metabolism, persistent and controlled delivery, and reduction of undesirable adverse effects [1,2]. However, the widespread use of this system is restricted to a few drugs only, because of the impermeable nature of the stratum corneum (SC), the outermost layer of the skin [3,4]. To overcome these limitations, many micro-structured fluid systems, including microemulsions, nanoparticles, and permeation enhancers containing other vesicles have been investigated to improve drug delivery efficiency via disrupting and modifying the regular arrangement of the corneocytes of the SC [5,6]. Among these fluid systems, attention has been focused on microemulsions (MEs), usually consisting of water, oil, and surfactant, which are useful colloidal nano-carriers for a TDDS, owing to their thermodynamic stability, high drug-loading capacity, and very low surface tension [7,8]. However, the conventional water-in-oil (W/O) and oil-in-water (O/W) ME systems are not suitable for drugs that are insoluble or sparingly soluble in water and most organic solvents [8–10].

Ionic liquids (ILs) are organic salts consisting of organic cations and inorganic and/or organic anions that have melting points below 100 °C. Because of their various unique physicochemical properties, ILs have become important in diverse scientific and technological arenas, especially used in ME systems in all the phases, formed by altering the water, oil, and surfactants [8,9,11]. The first IL-in-oil (IL/O) ME was developed by Moniruzzaman et al., in which the core aqueous phase was replaced by a hydrophilic IL, dimethylimidazolium dimethylphosphate ([C1mim][DMP]), which has emerged as a nano-carrier with great potential in the field of TDDSs for its excellent solubilizing capacity of sparingly soluble drugs [8,9]. Later other researchers have also used imidazolium ILs as the polar phase to increase the solubility and permeability of drugs [11,12]. However, most of the IL/O MEs were prepared with polyoxyethylene sorbitan monooleate (Tween-80), sorbitan laurate (Span-20), and other conventional non-ionic surfactants that require a high amount of surfactant for drug loading, which reduces the permeability [13], and increases the toxicity of the MEs [14]. Recently, surface-active ILs (SAILs) have been introduced as ILs, which act as surfactants to increase the physico-thermal stability [15,16] and the permeability of MEs [11].

Despite having various advantages, the use of ILs is limited because of their high toxicity and low biocompatibility and degradability [17]. In fact, most of the ILs used in previous studies, including imidazolium, pyridinium, and quinolinium cations and high strength inorganic acid anion-based ILs cannot be used in clinical applications because of their high toxicity and low biocompatibility [18,19]. Interestingly, ILs containing choline and amino acid esters as cations and organic acid anions (e.g., acetate, phosphate, and carboxylate) are non-toxic, biocompatible, and biodegradable [10,18,20]. It has been reported that ILs containing choline as the cation were less toxic compared with ILs containing imidazolium cations [21,22]. ILs containing choline as the cation and carboxylic acid as the anion are generally regarded as safe (GRAS), and considered to be non-toxic and biodegradable because of their biological sources [23–25].

In this study, we prepared IL/O MEs, using biocompatible ILs as replacements for imidazolium-based ILs, for the transdermal delivery of acyclovir (ACV), a model sparingly soluble antiviral drug. We selected choline formate ([Ch][For]), choline lactate ([Ch][Lac]), and choline propionate ([Ch][Pro]) ILs as the non-aqueous polar phases in the core of the MEs. In addition, a long chain (C_{18}) fatty acid SAIL, choline oleate ([Ch][Ole]) was incorporated as the surfactant in combination with Span-20 in a continuous oil phase of isopropyl myristate (IPM). The ILs and SAIL were selected because of their biocompatibility and negligible toxicity [24–26]. The solubility of ACV in the ILs and the IL/O MEs, the hydrodynamic size and size distribution of the ME droplets, and the stability of the MEs were studied. In addition, in vitro drug permeation into and across the skin was investigated using Yucatan micro pig (YMP) skin. Finally, a cytotoxicity study of the ILs and the IL/O MEs was performed using a three-dimensional cultured human epidermis model (LabCyte EPI-MODEL) and histological analysis.

2. Materials and Methods

2.1. Materials

ACV and IPM were obtained from Tokyo Chemical Industries Co. Ltd. (Tokyo, Japan). Choline chloride ([Ch][Cl]), silver oxide (Ag_2O), lactic acid, oleic acid, methanol, and acetonitrile were purchased from Wako Pure Chemical Industries Ltd. (Osaka, Japan). Formic acid and propionic acid were obtained from Kishida Chemical Co., Ltd., (Osaka, Japan). Tween-80 and Span-20 were procured from Sigma–Aldrich Chemical Co., (St. Louis, MO, USA). The skin of a female Yucatan micropig (YMP) was received from Charles River Japan Inc., Yokohama, Japan.

The three-dimensional cultured human epidermis model (LabCyte EPI-Model 12) was supplied by Japan Tissue Engineering Co., Ltd., Gamagori Miyakitadori, Aichi, Japan. 3-[4,5-Dimethylthiazol-2-yl]-2,5-diphenyltetrazolium bromide (MTT) was obtained from Dojindo Molecular Technologies, Inc., Kumamoto, Japan. All other chemicals and solvents used in the experiments were of analytical grade.

2.2. Synthesis of ILs

Biocompatible ILs ([Ch][For], [Ch][Lac], and [Ch][Pro]) and a SAIL ([Ch][Ole]) were selected based on their good toxicity and degradation profiles. The ILs and SAIL were synthesized using a two-step metathesis reaction following an established procedure with slight modification [24,26]. In the first step, choline hydroxide ([Ch][OH]) was synthesized by mixing a predetermined amount of choline chloride ([Ch][Cl]) and an excess amount of Ag_2O in Milli-Q water (Milli-Q) at room temperature for 2 h. Excess Ag_2O was removed from the reaction medium by centrifugation and filtration to obtain [Ch][OH]. In the second step, freshly prepared [Ch][OH] was neutralized with an equimolar aqueous solution of carboxylic acid (formic, lactic, or propionic acid) by continuous stirring at room temperature for 24 h (Scheme S1). As oleic acid was not soluble in Milli-Q, the neutralization of [Ch][OH] and oleic acid was performed in methanol instead of Milli-Q [23,26]. The solvent was evaporated using a rotary evaporator (EYELA, NVC-2200, Bohemia, NY, USA) at 40 °C. Finally, the synthesized ILs were freeze-dried for 48 h to evaporate the remaining solvents completely and the reaction yields were ≥85%. The synthesis of the ILs was confirmed by characterization using ^{1}H-NMR spectroscopy (JEOL Delta, ECZS NMR spectrometer, 400 MHz, Tokyo, Japan). The water content of all synthesized ILs was determined by Karl Fischer (KF) titration.

2.3. Solubility of ACV in [Ch][CA] ILs

The solubility of ACV in three choline carboxylic acid ILs ([Ch][CA]: [Ch][For]; [Ch][Lac]; and [Ch][Pro]) was determined by adding an excess amount of ACV to the ILs followed by continuous stirring for 24 h at 25 °C. Then, the undissolved ACV was removed by centrifugation followed by filtration using a syringe-driven filter (Millipore, 0.45 μm diameter). IPM and Milli-Q were used as a comparative control instead of the ILs. Finally, the filtrates were analyzed by ultraviolet-visible (UV-vis) spectrophotometry at 252 nm with suitable dilution in methanol to determine the ACV concentration according to a literature procedure [9].

2.4. Phase Behavior Studies of IL/S/Co_{mix}/IPM Systems: Preparation of IL/O MEs

A phase behavior study was performed prior to ME formation. First, the miscibility of the [Ch][CA] ILs was checked in the [Ch][Ole]/Span-20/IPM systems, where [Ch][Ole] SAIL and Span-20 were used as a surfactant (S) and co-surfactant (Co), respectively. Briefly, S and Co were blended in different weight ratios (*w/w*) (1:0, 3:1, 2:1, 3:2, 1:1, 2:3, 1:3, and 0:1). Then, a 15 wt.% of the S and Co mixture (S/Co_{mix}) was added to an appropriate amount of IPM, and the mixture was vigorously vortexed to obtain a clear and optically transparent homogeneous solution. Finally, [Ch][CA] ILs were added dropwise individually with continuous stirring until the final mixture turned turbid. The experiment was carried out at room temperature. Then, phase behavior studies were performed as stated above at selected weight ratios (2:1, 3:2, 1:1, 2:3, and 1:3) of S and Co, where the total S/Co_{mix} was maintained at 5 to 75 wt.%. For [Ch][For] and [Ch][Lac] ILs, the experiment was performed at a 2:1 ratio of S/Co_{mix} only. In addition, [Ch][Ole], Tween-80, and Span-20 were also mixed at a 1:1:1 weight ratio where [Ch][Pro] IL was used to carry out further processes.

Finally, IL/O MEs (IL/S/Co_{mix}/IPM) were prepared as stated above, where ILs, S/Co_{mix}, and IPM were maintained at 3, 15, and 82 wt.%, respectively. A water-in-oil (W/O) ME was also prepared where 3 wt.% Milli-Q was added instead of the IL, using the latter process, as shown in Table 1.

Table 1. Contents of the microemulsion (ME) formulations [a].

Formulations	ILs	Surfactant: Co-Surfactant (Weight Ratio)		
		Surfactant		Co-Surfactant
		[Ch][Ole]	Tween-80	Span-20
ME1	[Ch][Pro]	2	-	1
ME2	[Ch][Pro]	3	-	2
ME3	[Ch][Pro]	1	-	1
ME4	[Ch][Pro]	2	-	3
ME5	[Ch][Pro]	1	-	3
ME6	[Ch][Pro]	1	1	1
ME7	[Ch][For]	2	-	1
ME8	[Ch][Lac]	2	-	1
ME9 [b]	Milli-Q	-	2	1

[a] MEs were prepared with overall 15 wt.% S/Co_{mix} and 3 wt.% [Ch][CA] ionic liquid (IL) in isopropyl myristate (IPM). [b] ME9 was set as control where 3 wt.% Milli-Q was used as a replacement for [Ch][CA] IL.

2.5. *Viscosity, Density, and pH of the ILs and IL/O MEs*

The viscosity, density, and pH of each [Ch][CA] IL and IL/O ME formulation were measured at 25 °C using an automated micro-viscometer (Anton Paar Micro-viscometer, c, 2000M/ME, Graz, Austria), micro-densitometer (Anton Paar Density Meter, DMA 35N, Graz, Austria), and pH meter (TOA, HM-30R), respectively. The viscosity of the tested samples was evaluated considering the rolling time of the ball in a sample-filled glass capillary.

2.6. *Drug Loading Capacity of the IL/O MEs*

To determine the maximum drug loading capacity of the MEs, an excess amount of ACV was added to the ME formulations. The ACV-loaded MEs were stirred for 24 h at room temperature. The unloaded ACV was removed using a centrifugation and filtration method. The amount of drug in the subsequent clear filtrate was measured using a UV spectrophotometer as described in Section 2.3.

2.7. *Particle Size Determination*

The hydrodynamic size and polydispersity index (PDI) of drug-loaded/unloaded MEs were determined by dynamic light scattering (DLS: Zetasizer Nano ZS, Malvern Instruments, Worcestershire, United Kingdom). All the tested samples were equilibrated for more than 4 h before starting measurements, and there was no visible macroscopic heterogeneity. Samples were equilibrated for approximately 10 min before collecting data. The average diameters of the tested samples were calculated using five replicated experiments.

2.8. *Stability of the ME Formulations*

The stability of the drug-loaded MEs was investigated for two months considering storage time and storage temperature. The stability was determined by measuring the droplet size using DLS and visual inspection at regular intervals. In addition, the physical stability of the MEs was determined by centrifugation for 30 min at 15,000 rpm. The chemical stability was also examined by measuring the drug degradation extent and encapsulation efficiency.

2.9. Skin Permeation Studies

In vitro drug permeation was investigated on YMP full thickness skin using hand-made Franz diffusion cells (10 mm diameter), consisting of donor and receiver compartments. The prepared skins (2 cm × 2 cm) were soaked in D-PBS solution for 1 h prior to the permeation experiment. The skins were then clamped on the Franz cell with the SC facing up to the donor compartment and the dermis contacting with the receiver phase (D-PBS, pH 7.4). Then 0.5 mL of each of the ACV-loaded formulations (IL/O MEs and controls) were applied on the donor compartment. The receiver compartment was maintained thermostatically at 32.5 ± 0.1 °C using a circulating water bath (NTT-20S, Tokyo Rikakikai Co. Ltd., Tokyo, Japan) and magnetically stirred at 500 rpm during the entire experiment. After a fixed interval, 0.5 mL of the receiver solution was withdrawn to determine the transdermal drug delivery content (permeated across the skin), while an equal amount of fresh D-PBS was added to maintain a constant volume (5 mL) of the receiver solution. After 48 h, the skins were unclamped and washed with 0.1 M HCL three times to remove the tested formulations completely from the skin surface. Finally, the treated skins were processed according to our previous report [27] to estimate the topical drug delivery content (penetrated into the skin). The concentration of ACV was determined using HPLC with a Shiseido CAPCELL PAK C18 MG (4.6 mm × 250 mm) column using the United States Pharmacopeia (USP) method according to a previous report [28], where the mobile phase was 0.02 M glacial acetic acid with elution at a flow rate 1.5 mL/min, and the injected volume of sample was 100 µL. The concentration range of the standard curves was 0–25 µg/mL, and the squared correlation coefficient of the standard curve was more than 0.99 ($R^2 > 0.99$).

2.10. Calculation of Skin Permeation Parameters

The cumulative amount of ACV (Q_h, µg/cm^2) that permeated across the skin was plotted as a function of time, in order to determine the various permeability parameters, where the transdermal flux (J, µg/cm^2/h) was calculated as the slope. The permeability coefficient (K_P, cm/h), was measured using the following equation: $K_P = J/C_d$, where C_d (µg/mL) was the concentration of drug in the donor phase. Lag time (t_L, h) was the intercept of the X-axis. The diffusion coefficient (D, cm^2/h) was calculated from the lag time by the equation, $D = l2/6t_L$, where l (cm) was the thickness of the skin. The skin partition coefficient, (K_{skin}), was calculated from the following equation: $K_{skin} = (Jl)/(DC_d)$.

2.11. Impact of the IL/O MEs on the Skin Barrier Properties

The skin samples were prepared according to a previous report with some modifications [29]. Full thickness YMP skin was thawed and allowed to stand for 1 h at room temperature. Then, the fat portion of the skin was cut off to make it moisture free, and the skin was incubated at 60 °C for 1–2 min to loosen the epidermis. After pulling out the epidermis, the epidermis was floated on a 0.25% trypsin and 1 mM EDTA solution for 24 h at room temperature. The SC side of the epidermis was faced up during floating. Then the SC was isolated from the epidermis and washed with water and allowed to dry for 24 h at room temperature. After cutting into the desired size, the SC was soaked in the test MEs for 30 min at room temperature. Then, the SC was withdrawn from the test samples and washed thoroughly with 20% ethanol and allowed to dry for 1 h. Finally, the treated SC was analyzed by Fourier transform infrared spectroscopy (FTIR) and compared with untreated skin (control).

2.12. Cytotoxicity Evaluation of ILs and IL/O MEs

The in vitro cytotoxicity study was performed using a three-dimensional cultured human epidermis model (LabCyte EPI-Model 12, J-TEC, Japan) according to a previous report with some modifications [9]. Briefly, the tissues were cultured into 24-well plates (BD Biosciences, San Jose, CA, USA) with assay medium (0.5 mL) and were incubated for 24 h at 37 °C in a 5% CO_2 humidified environment. Then, 25 μL of the test formulations were applied into each well on the tissue surface, D-PBS and commercial IL [C1mim][DMP] treated samples were used as negative and positive controls, respectively. The cultures were then incubated for 24 h (37 °C, 5% CO_2). After that, the tissues were withdrawn from the culture media and washed 15 times with D-PBS carefully to remove any remaining formulation from the tissue surface. Then, 0.5 mL of freshly prepared MTT solution (0.5 mg/mL) was added to each well, and the wells were incubated for 3 h (37 °C, 5% CO_2). The tissues were then immersed fully into 0.5 mL of propan-2-ol containing micro-tubes and allowed to stand in a refrigerator for 48 h after covering with aluminum foil. Finally, 100 μL of the extracted solutions was transferred into a 96-well plate for measuring the optical density at 570 and 650 nm (as a reference absorbance) using a microplate reader (iMARK, Bio-Rad, Tokyo, Japan), where propon-2-ol was used as a blank. The cell viability was calculated as the percentage relative to the negative control, D-PBS.

2.13. Histological Study

The dermal safety of the tested IL/O MEs was investigated on YMP skin according to a previous report [9]. First, the desired size (2 cm × 2 cm) skins were treated with the IL/O MEs or D-PBS (1.0 mL) for 24 h where D-PBS was used as the control. The skin samples were then immersed into Histo Prep compound (Fisher Scientific, Branchburg, NJ, USA) at −80 °C followed by sectioning using a cryostat microtome (CM1510; Leica, Wetzlar, Germany) and placed on glass slides. The slides were then stained with hematoxylin and eosin solution (Muto Pure Chemicals Co. Ltd, Tokyo, Japan). Finally, the specimens were investigated under a high-powered light microscope (BZ-9000 BIOREVO, Keyence Corp., Itasca, IL, USA).

2.14. Statistical Data Analysis

The data are given as the mean ± standard deviation (SD). The comparisons between more than two groups were performed by two-way ANOVA analysis for multiple comparison tests using GraphPad Prism software (Version 6.05). The differences were considered significant at $p < 0.05$.

3. Results and Discussion

3.1. The Solubility of ACV in the [Ch][CA] ILs and Relationship to the Physical Properties of the ILs

The ILs and SAIL used in this study were selected by considering the two important factors, biocompatibility and toxicity. From a structure-activity relationship point of view, choline is the most suitable candidate as a cation, as choline can be derived from various natural sources and has multiple biological functionalities [18,30]. Choline is also known as a source of macronutrients [18,24]. Carboxylic acids are GRAS and have been widely used as pharmaceutical solvents for a long time [25,31,32]. Therefore, three [Ch][CA] ILs consisting of choline as the cation and a carboxylic acid (formic, lactic, or propionic acid) as the anion [Ch][For], [Ch][Lac], and [Ch][Pro], respectively, and one SAIL consisting of choline as the cation and a long chain (C18) fatty acid as the anion [Ch][Ole], were considered to be safe and biocompatible for further study in a TDDS. [Ch][For], [Ch][Lac], and [Ch][Pro] are hydrophilic in nature and intended for use to investigate the solubilization capacity of the sparingly soluble drug, ACV and [Ch][Ole] is known to act as a surfactant [26].

Three fundamental properties of the [Ch][CA] ILs, the viscosity, density, and pH were measured, as shown in Table 2. The viscosity and density of the three [Ch][CA] ILs were significantly different with a good agreement to previous reports [24]. The measured pH values varied from 5.5 to 7.6. It has been reported that the physical properties of synthesized ILs are directly influenced by the structure, symmetry, and alkyl chain length of the carboxylic acid [24]. However, the actual relationship of the physical properties of ILs with the nature of the cation/anions could not be established in this study.

Table 2. The solubility of acyclovir (ACV) in the [Ch][CA] ILs [a] at 25 °C and the effects of the viscosity, density, pH, and anionic domain on IL-mediated dissolution [b].

IL or Solvent	Anionic Structure R—C(=O)OH	Solubility of ACV (mg/mL)	ρ (g/cm^3)	η (m Pa s)	pH
[Ch][For]	R = -H	203 ± 12 ****	1.12 ± 0.03	124.7 ± 7	5.5
[Ch][Lac]	R = -C_2H_5O	208 ± 15 ****	1.15 ± 0.02	897.2 ± 27	7.6
[Ch][Pro]	R = -C_2H_5	278 ± 18 **,****	1.07 ± 0.02 †	309.5 ± 12 ####	6.2
IPM	-	0.03 ± 0.01	-	-	-
Milli-Q	-	0.41 ± 0.08	-	-	-

[a] The solubility of ACV in water and IPM is also given. [b] Data are shown as mean ± SD (n = 3). **** compared with Milli-Q and IPM, $p < 0.0001$; ** compared with [Ch][For] and [Ch][Lac], $p < 0.01$; † compared with [Ch][For] and [Ch][Lac, $p < 0.05$; #### compared with [Ch][For] and [Ch][Lac], $p < 0.0001$ using Sidak's multiple comparison test.

The maximum solubility of ACV in the ILs was 203, 208, and 278 mg/mL for [Ch][For], [Ch][Lac], and [Ch][Pro], respectively, which was significantly higher compared with Milli-Q and IPM (Table 2). As the ILs were hydrophilic in nature with strong H-bond accepting anions, ACV may be dissolved in these ILs through the formation of H-bonds, van der Waals forces, or π–π interactions between the polar groups of the drug and the IL anions [8]. It has been reported that the solubility of a drug depends on several factors, i.e., H-bond accepting ability, density, viscosity, and the alkyl/aromatic side chains of the IL anion [33,34]. The solubility of ACV was higher in [Ch][Pro] owing to its low viscosity, and the small charge-localized anion. On the other hand, [Ch][For] formed interionic hydrogen bonds, having a very small charge-localized anion and [Ch][Lac] formed intramolecular hydrogen bonds, having an extra –OH group, resulting in less hydrogen bonding ability with ACV and less ACV solubilizing capacity [33].

3.2. Phase Behavior Studies of IL/S/Co_{mix}/IPM Systems: Preparation of IL/O MEs

A phase behavior study is very important for selecting the optimum composition of the MEs. Prior to the phase behavior study, a miscibility study of the ILs in a S/Co_{mix}/IPM system (consisting of 15 wt.% S/Co_{mix} at different S/Co weight ratios) was performed, and it was found that a higher [Ch][Ole] content in the S/Co_{mix} was favorable for the miscibility of all ILs in S/Co_{mix}/IPM system, as shown in Figure S1. However, all the ILs were immiscible at 0:1, 1:0, and 3:1 S/Co ratios in S/Co_{mix}/IPM and the miscibility of [Ch][For] and [Ch][Lac] was very low at all S/Co ratios, except 2:1. Therefore, the phase behavior study was performed at 2:1, 3:2; 1:1, 2:3, and 1:3 S/Co ratios for [Ch][Pro] (Figure 1 and Figure S2), whereas only a 2:1 ratio was used for [Ch][For] and [Ch][Lac] (Figure S3). In the case of [Ch][Pro], it was found that the area of the single phase or ME forming regions varied with the S/Co ratio and the trend was 2:1 > 3:2 > 1:1 > 2:3 > 1.3 (Figure 1 and Figure S2). As the ILs are hydrophilic in nature and immiscible with IPM, they must be located in the core of the micelle owing to the hydrogen bond formation between the –OH groups of S/Co and the anions of the ILs, and the strong electrostatic interaction between the positive head group of Ch][Ole] and the anion of the ILs [8]. Because of the strong electrostatic interaction between the head group of [Ch][Ole] and the anion of the ILs, a higher content of [Ch][Ole] in the S/Co_{mix} favored ME formation [9,35]. To compare the surface activity of [Ch][Ole] and Tween-80, the phase behavior of a ME consisting of [Ch][Ole]/Tween-80/Span-20 at a 1:1:1 weight ratio was studied (Figure S2), and it was found that replacing [Ch][Ole] by the same amount of Tween-80 caused the ME to lose IL holding capacity. This result indicated that [Ch][Ole] has a higher surface activity than Tween-80, owing to the higher electrostatic interaction between the head group of [Ch][Ole] and the anion of the polar IL [9]. In addition, compared with conventional surfactants, a much lower percentage of S/Co_{mix} was required to solubilize a large amount of ILs using [Ch][Ole]. It has been reported that MEs required a comparatively reduced amount of surfactant with a two or more surfactant mixture than with a single surfactant [36]. This interesting finding can be explained in terms of favorable interfacial properties (e.g., rigidity and polarity) provided by the mixture of the two surfactants [8]. Among the three ILs, [Ch][Pro] had the maximum ME formation capacity, which was indicated by the larger single-phase area in the phase diagram (Figure S3) and this greater capacity was because of the greater hydrogen bonding ability of [Ch][Pro] [33].

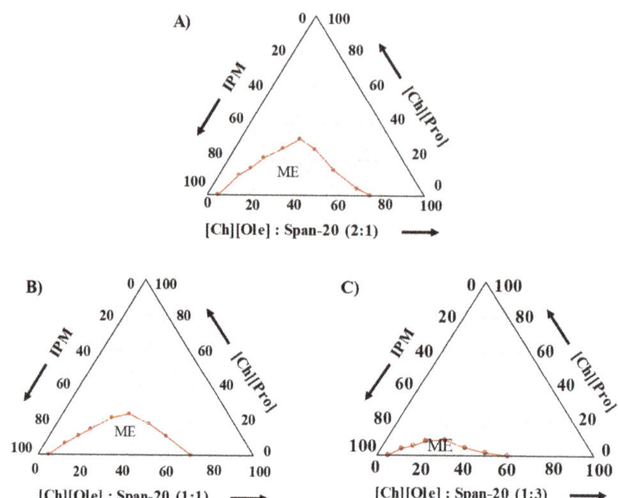

Figure 1. Phase behavior studies of IL/S/Co_{mix}/IPM MEs consisting of [Ch][Pro] with varying weight ratios of S/Co (**A**) 2:1, (**B**) 1:1, and (**C**) 1:3 at 25 °C.

Finally, we prepared MEs with optimum compositions (Table 1). Though a larger amount of S/Co_{mix} in the MEs could retain a larger amount of ILs, resulting in a larger amount of drug that could be loaded, this reduced the permeability [13] and increased the toxicity [14]. Therefore, we selected a comparatively lower content of IL and S/Co_{mix}, and a higher content of IPM in the ME formulations in this study. ME6 was prepared to compare the surface activity, drug loading capacity, permeability, and toxicity of [Ch][Ole] and Tween-80.

3.3. Density and Viscosity of the MEs

The density of the MEs decreased with increasing [Ch][Ole] content in the S/Co_{mix} because of the lower density of [Ch][Ole] (0.98 g/cm^3) compared with Span-20 (1.032 g/cm^3), but the differences were not significant among them. The viscosity of the MEs varied significantly with the S/Co ratio and increased with increasing [Ch][Ole] content in the S/Co_{mix}, whereas the viscosity did not depend on the ILs, Table S1.

3.4. Particle Size Determination

The hydrodynamic size and PDI of the MEs were determined by DLS. The particle size variation of the MEs (consisting of 15 wt.% S/Co_{mix} at a 2:1 weight ratio), with varying R values (molar ratios of IL per S/Co_{mix}) were studied, and it was found that the particle size increased with increasing R values, which confirmed that the IL was located in the hydrophilic micelle core (Figure S4). When the IL was added, the additional IL entered the core of the ME. To cover the additional IL, the surfactant aggregates were expanded, resulting in a larger particle diameter of the MEs [8]. In addition, by plotting particle size as a function of R, it was found that the particle size was an almost linear function of R, as shown in Figure S5, and according to the swelling law of MEs, this indicated spherical ME droplets [37]. Moreover, the particle size of the MEs was also studied at constant R values (by varying the surfactant and IL), and it was found that with increasing surfactant content, the number of particles was also increased while keeping a constant particle size (Figure S6) [8].

The particle size of the MEs was determined with varying S/Co weight ratios, and it was found that the particle size of the MEs varied from 17.7 to 31.3 nm (less than 100 nm), which indicated satisfactory MEs [38], and the particle size increased with increasing [Ch][Ole] and/or decreasing Span-20 content (Figure 2A). This trend can be explained based on the content of the individual surfactants in the S/Co$_{mix}$. With higher [Ch][Ole] content in the S/Co$_{mix}$, the head group of [Ch][Ole] might face steric hindrance with the cation of the polar IL, and consequently, the particle size increased. On the other hand, with a higher content of Span-20, the ME has the ability to form smaller reverse micelles in the organic media by marked surface bending of the large hydrophobic chains [10,39]. In addition, there is a positive correlation between the viscosity and the particle size of MEs [9,10]. The particle size of the [Ch][Pro]-based ME was smaller than [Ch][For] and [Ch][Lac] at the same S/Co ratio owing to the stronger hydrogen bonding ability of [Ch][Pro] (Figure 2A).

The size and size distribution of ACV-loaded (2 mg/mL) MEs with varying S/Co ratios were also studied to investigate the effect of drug loading on the droplet size, and it was found that the particle size decreased compared with the drug-free MEs for all S/Co ratios studied (Figure S7). In further investigations, the particle size of ME1, loaded with different concentrations (0, 1, 3, and 5 mg/mL) of ACV was also studied, and it was found that the particle size decreased from 31 to 22 nm with increasing drug concentration (Figure 2B), and this trend was in good agreement with previous literature [9,40]. There may be two possible reasons for this effect, firstly, at higher drug concentrations, a certain amount of drug might be deposited into the interphase of the ME, and thus, reduce the particle size by acting as an emulsifying agent and secondly, the deposited drug at the interphase could reduce the surfactant movement and consequently reduce the particle size [9]. The small values of the PDI (<0.3) indicated the homogeneity of the prepared MEs [38].

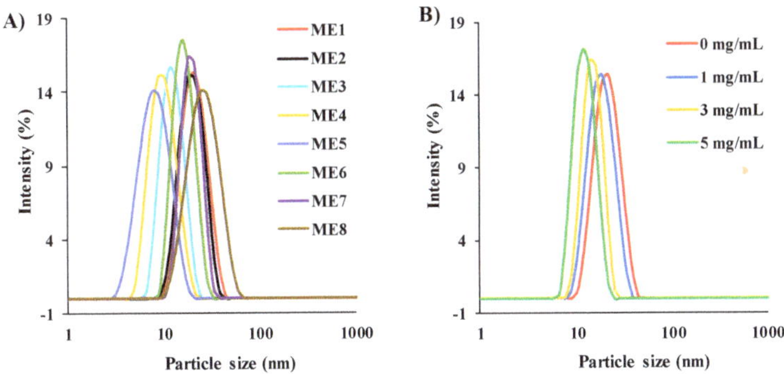

Figure 2. The size and size distribution of (**A**) drug-free MEs with varying S/Co weight ratios and (**B**) drug-loaded (0–5 mg/mL) ME1 at 25 °C.

3.5. Drug Loading Capacity of the IL/O MEs

The drug loading capacity of the IL/O MEs was estimated to assess them as a vehicle for the delivery of a sparingly soluble drug. Though the ACV-loading capacity of the IL-free S/Co$_{mix}$/IPM system was very low (0.15 mg/mL for S/Co = 2:1, 15 wt.%), the capacity was increased dramatically by incorporating IL into this system (7.7 mg/mL for ME1). The ACV-loading capacity was decreased with decreasing [Ch][Ole] content in the S/Co$_{mix}$, and it was significantly decreased when [Ch][Ole] < Span-20 in the S/Co$_{mix}$, as shown in Figure 3. The ACV-loading capacity also depended on the type of IL. Comparing ME1, ME7, and ME8, it was found that ME1 had a significantly higher loading capacity than ME7 or ME8, owing to the higher ACV solubilizing capacity of [Ch][Pro]. It has been reported that the drug loading capacity of IL/O MEs highly depends on the categories of IL [10]. In addition, comparing ME1 and ME6, it can be seen that [Ch][Ole] had more influence on the drug loading in this system than Tween-80. Among all the formulations, ME1 showed the highest drug loading capacity because of its larger stable interface compared with the other MEs [9,10,41]. Moreover, the relative solubility of the drug in [Ch][Ole], Tween-80, and Span-20 would have a contribution to the ability of a given ME to entrap the drug [9].

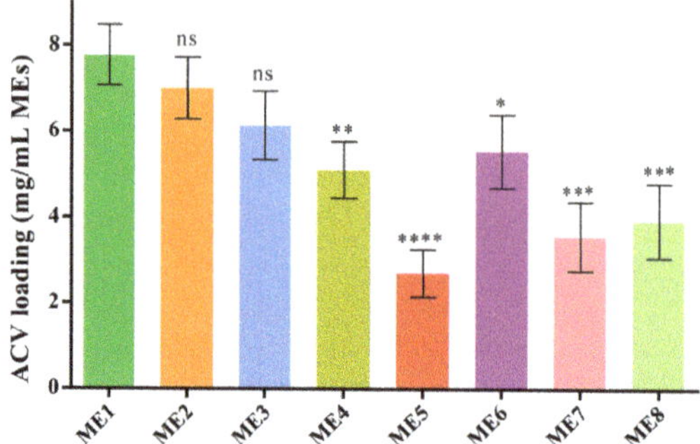

Figure 3. ACV-loading capacity of MEs at 25 °C; (mean ± SD, $n = 3$, ns: not significant, * $p < 0.05$, ** $p < 0.01$, *** $p < 0.001$, and **** $p < 0.0001$ using Dunnett's multiple comparison test.

3.6. Stability of the Drug-Loaded MEs

It is important to note that MEs need to be stable to be used as drug delivery carriers. To investigate the stability of the MEs, in this study we examined ACV (5 mg/mL) loaded MEs (ME1, ME2, ME3, ME4, and ME6) over two months at 25 °C. As the ACV-loading capacities of ME5, ME7, and ME8 were <5 mg/mL, they were not considered for stability and drug delivery experiments. No significant change was found in terms of clarity and phase separation observations, and the particle sizes of ME1 and ME2, during the entire observation time. The particle size of ME1 was increased slightly from 22 to 25 nm, which was not significant, as shown in Figure 4A. However, the particle sizes of ME3, ME4, and ME6 started to increase linearly from 30 days, and finally, the samples became turbid after 45 days, which confirmed the formation of stable MEs with a higher [Ch][Ole] content in the S/Co_{mix}. It has been reported that SAIL can increase the stability of MEs [15,16]. No physical instabilities (e.g., phase separation, phase inversion, aggregation, or cracking) of the MEs were found by centrifugation, which confirmed the physical stability and excellent drug encapsulation efficiency of the MEs [42]. In addition, ACV-loaded ME1 and ME2 were stored at different temperatures (4, 25, and 37 °C) for two months to assess the effect of storage temperature on stability. After several time intervals, the MEs were examined by visual inspection and particle size determination. No significant change was found for either ME after two months (Figure 4B for ME1), indicating a negligible impact of the storage temperature on the long-term stability of drug-loaded MEs. This stability can be explained in terms of the ionic ([Ch][Ole]) and non-ionic (Span-20) character of the surfactant. It has been reported that a mixture of ionic and non-ionic surfactants can form temperature-insensitive MEs owing to their synergistic effects [36]. The chemical stability of ACV-loaded ME1 and ME2 formulations was investigated using HPLC, and it was found that the MEs showed excellent encapsulation efficiency of ACV. After two months, the encapsulation efficiency of both MEs (contained 5 mg/mL ACV initially) was ≥98%, indicating no degradation (Figure S8).

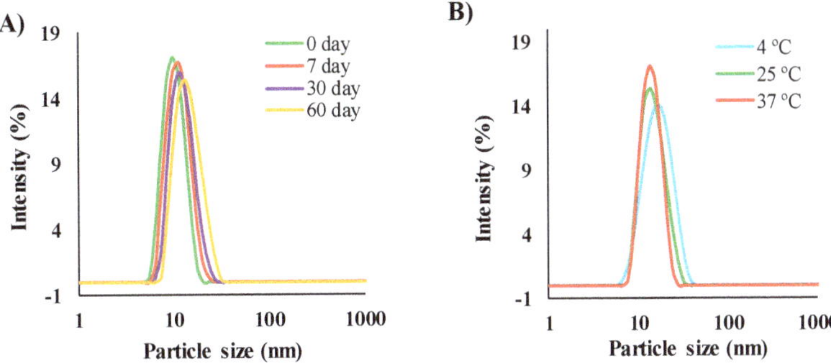

Figure 4. The size and size distribution of ACV-loaded ME1 (**A**) effect of storage time at 25 °C and (**B**) effect of storage temperature after two months.

3.7. Skin Permeation Studies

In vitro drug permeation studies were performed using YMP skin owing to its similar clinical, structural, and immuno-histochemical features to human skin [9]. First, the topical and transdermal delivery of ACV from IL/O MEs was investigated and compared with other formulations (Figure 5A). The topical delivery of ACV from IPM, S/Co_{mix}/IPM, and W/O MEs was very low, while the transdermal delivery was below the detection limit. Interestingly, compared with the other formulations, the IL/O ME demonstrated significantly enhanced topical and transdermal delivery with values of 36.47 and 45.05 μg/cm^2, respectively. This dramatically enhanced permeation using the IL/O MEs was found because of their high drug solubilizing capacity and promising drug conveyances technique. Generally, drugs are administrated into the skin in a solubilized state. A large amount of drug was loaded into the core of the IL/O ME solubilized by IL, which could act as a drug reservoir and provide a greater concentration gradient to the skin [9]. Whereas, IPM (a potential enhancer) disrupts the barrier function of the skin, which facilitated to enter the nano-sized drug-loaded IL droplets into the skin [11]. Nonetheless, ACV was solubilized state in IL, but as it is hydrophilic in nature, IL alone could not deliver ACV due to of the strong hydrophobic barrier functions of the skin [9]. On the other hand, though IPM, S/Comix/IPM, and W/O MEs disrupt the barrier function as they contain IPM, ACV could not permeate across the skin from these formulations because ACV was suspended in these systems, which probably obstructed the access of ACV to the skin [9,28].

It has been reported that the molar ratio of individual surfactants can influence drug delivery by controlling the physicochemical properties of the MEs [43]. Therefore, the delivery of ACV from various MEs with varying S/Co ratios was studied. From the cumulative permeation profiles (Figure 5B), it can be seen that ME1 enhanced the transdermal delivery significantly compared with the other MEs. Other permeation parameters, including transdermal flux, permeability coefficient, diffusion coefficient, and skin partition coefficient were determined from the cumulative permeation profile. It was found that all these parameters were increased with increasing [Ch][Ole] content in the MEs, as shown in Table 3, indicating that higher [Ch][Ole] content was favored for transdermal delivery. In fact, the transdermal delivery of drug mainly depends on the transdermal flux and permeation coefficient. The highest transdermal flux (1.43 μg/cm^2/h) and permeation coefficient (2.86×10^{-4} cm/h) were both found for ME1, because of the higher skin partition and diffusion coefficients (indicating better solvent distribution ability into the deeper layers of the skin) resulting in ME1 having the highest transdermal delivery of ACV [11,44].

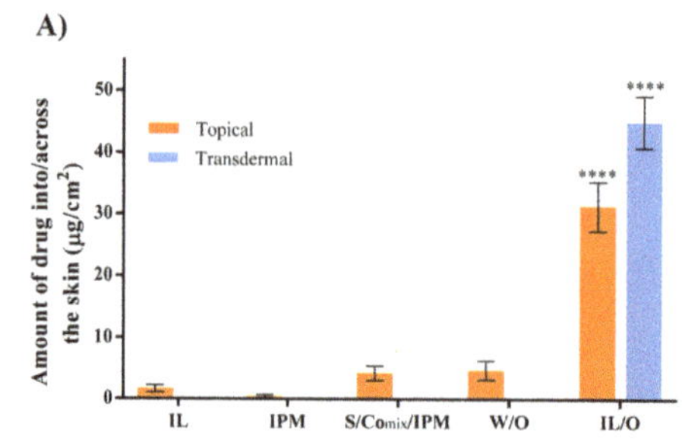

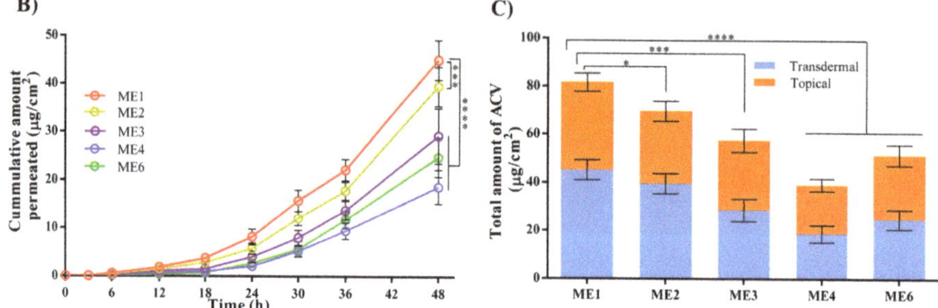

Figure 5. (**A**) Topical and transdermal delivery of ACV from various drug carriers after 48 h, where S/Co_{mix}/IPM: 15 wt.% S/Co_{mix} at a 2:1 ratio in IPM. W/O: ME9. IL/O: ME1; (**B**) transdermal permeation profile of ACV from various IL/O MEs with varying S/Co ratios; (**C**) the total (topical + transdermal) delivery of ACV from various IL/O MEs with varying S/Co ratios after 48 h; (mean ± SD, $n = 3$, * $p < 0.05$, *** $p < 0.001$, and **** $p < 0.0001$ using Dunnett's multiple comparison test. All the drug carriers contained 5 mg/mL ACV.

Table 3. Effect of [Ch][Ole] on permeation parameters. Data are shown as mean ± SD, n = 3.

Formu-Lations	Cumulative Amount, Q_{48h} (μg/cm²)	Transdermal Flux, J (μg/cm²/h)	Permeability Coefficient, K_P (×10⁻⁴ cm/h)	Diffusion Coefficient, D (×10⁻⁴ cm²/h)	Skin Partition Coefficient, K_{Skin}
ME1	45.05 ± 4.18	1.43 ± 0.13 ***,****	2.86 ± 0.24 ***,****	2.77 ± 0.22	0.18 ± 0.03
ME2	39.48 ± 4.14	1.21 ± 0.11	2.42 ± 0.21	2.68 ± 0.23	0.16 ± 0.04
ME3	28.53 ± 4.68	0.92 ± 0.11	1.85 ± 0.18	2.63 ± 0.25	0.12 ± 0.03
ME4	18.66 ± 3.50	0.62 ± 0.09	1.23 ± 0.15	2.58 ± 0.17	0.08 ± 0.02
ME6	24.78 ± 4.05	0.83 ± 0.1	1.66 ± 0.18	2.48 ± 0.22	0.12 ± 0.03

*** $p < 0.001$ compared with ME2 and **** $p < 0.0001$ compared with other MEs using Dunnett's multiple comparisons test.

In addition, the topical delivery of ACV was investigated. The total (topical and transdermal) delivery after 48 h is presented in Figure 5C. It was observed that, as for transdermal delivery, topical delivery was also favored using ME1, having a higher [Ch][Ole] content, resulting in the highest drug delivery. This result could be explained based on the higher interfacial area and stability of ME1. It has been reported that a larger stable interfacial area of ME droplets favors transdermal and topical delivery [9,45]. Comparing the transdermal delivery between ME1 and ME6, it was revealed that [Ch][Ole] significantly enhanced the permeability compared with Tween-80 (Table 3). In addition, by comparison with a previous report [9], (where a very low amount of ACV was permeated using a Tween-80/Span-20 surfactant-based ME, using the same experimental protocol), we can claim that the [Ch][Ole] has a significantly greater permeation enhancing ability compared with Tween-80. This enhanced ability can be explained in terms of the influence of the [Ch][Ole] on the skin modification being ionic in character [11]. Hence, the effect of IL/O MEs on the skin barrier properties was further studied.

As ME1 displayed the significantly higher transdermal flux and permeation coefficient (Table 3), and delivered the significantly higher amount of drug topically and transdermally (Figure 5C) than other MEs, could be the most suitable nano-carrier.

3.8. Impact of IL/O MEs on the Skin Barrier Properties

FTIR spectroscopy, which can provide deep insight into the molecular structure of the lipid matrix of the SC [46], was performed to assess the structural changes of the SC. To avoid the interference of ACV, drug-free MEs were applied, and the results were compared with an untreated sample (as control), as shown in Figure 6 and Table S2. All the treated samples produced some red shifts of the absorption peaks to higher wavenumber for both the lipid and keratin of the SC. For ME1, the C–H vibration peaks shifted to 2924.5 from 2920 cm^{-1} (asymmetric vibration), and 2854.5 from 2851 cm^{-1} (symmetric vibration), and the NH–C=O vibration peaks shifted to 1647.5 from 16440, and 1540.25 from 1538 cm^{-1}. These shifts are directly related to the molecular structure of the skin [46]. When the skin was treated with the MEs, the orthorhombic conformation of the lipids was transformed to a liquid crystalline conformation resulting in the CH_2 symmetric stretching vibration peak shifting to higher wavenumber. The shift of the NH–C=O vibration peaks to higher wavenumber revealed that the conformation of keratin was converted from an organized α-helical structure to a randomly coiled structure by treatment with the test systems. These shifts occurred because of the presence of lipophilic IPM in the test formulations, which disrupted the barrier function of the skin [11,44]. ME1 had a greater effect on the structural changes of the skin compared with the other MEs in the following order: ME1 > ME6 > ME9, which can be ascribed to the ionic character of the IL [11]. In ME1, the total ionic surfactant ([Ch][Ole]) was 10 wt.%, whereas it was only 5 wt.% in ME6, and ME9 was fully composed of non-ionic surfactants (Tween-80 and Span-20) and had no IL content. It has been reported that MEs containing an IL as surfactant disrupt the barrier function effectively because of their ionic character [11]. It has also been reported that an IL can alter the SC structure transiently by extracting the lipid from the skin [25]. The aforementioned results demonstrate that the IL/O MEs reduced the barrier properties of skin, leading to a high permeability coefficient and good drug delivery [11,47].

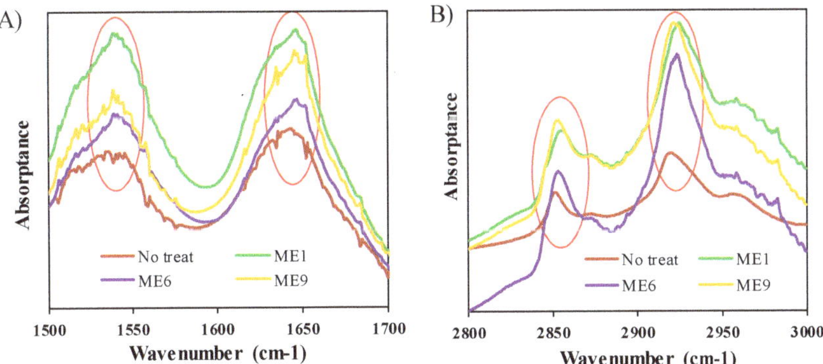

Figure 6. FTIR spectra of (**A**) keratin and (**B**) lipid of SC samples after treatment with different formulations.

3.9. Cytotoxicity Evaluation of ILs and IL/O MEs

In vitro cytotoxicity studies (skin irritation profiles) of the ILs and IL/O MEs were performed using a reconstructed human epidermal model (LabCyte EPI-MODEL-12). In this experiment, [Ch][Pro], ME1, ME6, and ME9 were selected to compare the relative toxicity with D-PBS (as a negative control) and [C1mim][DMP] (a commercial IL: IL_{com} as a positive control), as shown in Figure 7. The cell viability of all the MEs was above 92% compared with D-PBS, and the values were not significantly differed from D-PBS and IPM, which indicated that the prepared IL/O MEs were non-toxic. Comparing between ME1, ME6, and ME9, it was found that the toxicity profile of [Ch][Ole] was similar to Tween-80, which was in a good agreement with a previous report [26]. However, when [Ch][Pro] was used alone, the cell viability decreased to 48%, but the cell viability was below 15% for [C1mim][DMP], which suggested that [Ch][Pro] was less toxic than [C1mim][DMP] IL. The results were in a good agreement with published reports in which choline-based ILs demonstrated less irritation towards different cultured cell lines, such as human keratinocytes cell line (HaCat) [21], human embryonic kidney cell line (HEK-293) [18], and human epidermal keratinocytes-adult cell line (HEK-a) [25]. It has also been reported that imidazolium-based ILs are considered as toxic and less biodegradable, whereas choline-based ILs are regarded as safe, non-toxic and biocompatible [10,20]. This reduced toxicity can be attributed to the biocompatible sources of the cation and anion of [Ch][Pro], choline is used as a food additive and known to be non-toxic and biocompatible [18,25,48], and propionic acid is GRAS and used as preservative in food, cosmetic and pharmaceutical industries [31,32]. Our IL/O MEs, containing a low amount of IL, appears to be non-toxic and are potential carriers for TDDSs.

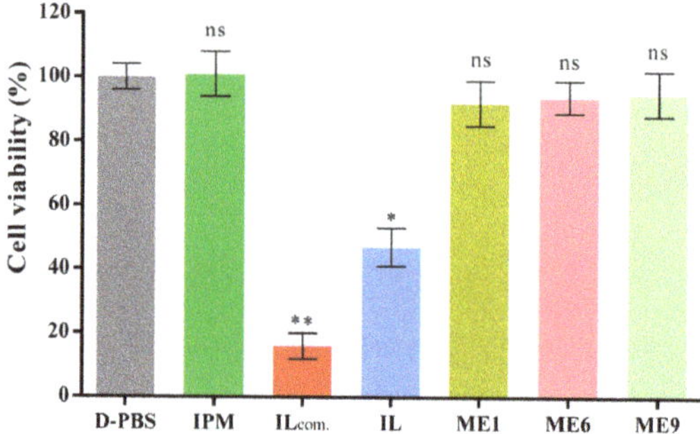

Figure 7. Cytotoxicity evaluation of ILs and MEs using reconstructed human epidermal model LabCyte EPI-MODEL-12 (mean ± SD, n = 3), ns: not significant, * $p < 0.05$ and ** $p < 0.01$ using Dunnett's multiple comparisons test.

3.10. Histological Study

IL/O MEs need to be safe and non-toxic to be used as transdermal carriers. According to the drug loading capacity, stability, and drug permeation studies, ME1 was the most suitable candidate for use as a transdermal carrier. Therefore, an in vitro histological study was performed to investigate the dermal safety of ME1. The ME1-treated skins were observed through a fluorescence microscope (20-fold magnification), and it was found that the structures of the SC, epidermis, and dermis of the skin were clearly visible and organized after treatment with ME1 for 24 h compared with control (D-PBS treated sample), as shown in Figure 8 which were in a good agreement with a published report [9]. Therefore, the IL/O MEs used in this study had no antagonistic effect on the skin, and could be a safe and promising nano-carrier for the transdermal delivery of sparingly soluble drugs.

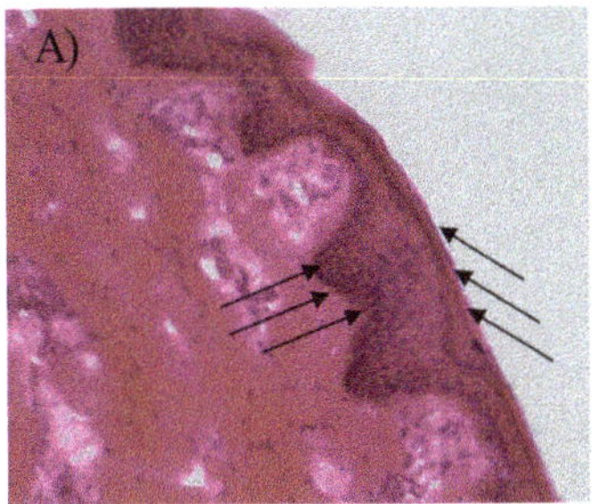

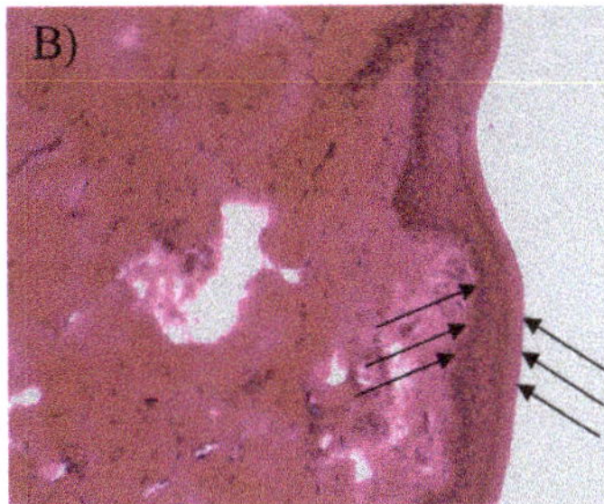

Figure 8. In vitro histopathological evaluation of YMP skin sections (×20) treated with (**A**) D-PBS and (**B**) ME1. The left direction arrows indicate the SC layer and the right direction arrows indicate the epidermis layer.

4. Conclusions

This study presents a novel IL/O ME which was developed using biocompatible ILs as the non-aqueous polar phase (core of the ME), as well as the surfactant, for an improved TDDS for the sparingly soluble drug, ACV. Preliminary results clearly indicated the optimum S/Co_{mix} in which to prepare a thermodynamically stable ME, with a high drug loading ability and enhanced drug permeability. FTIR investigations revealed that the enhanced drug permeation with the IL/O ME was because of a reduction of skin barrier function via modification and disruption of the regular arrangement of the corneocytes of the SC. If considering the drug loading capacity and skin permeation studies, the successful formation of an ME with [Ch][Pro] in the core as a non-aqueous polar phase could be attributed to the favorable interfacial properties provided by a blend of [Ch][Ole] and Span-20, compared with a blend of Tween-80 and Span-20. Moreover, in vitro skin irritation and histological tests confirmed that the prepared IL/O MEs were safe and non-toxic. Both the ILs, as well as the prepared IL/O MEs, were completely biocompatible, and are potential candidates for future applications in pharmaceutical formulations of sparingly soluble drugs, as well as proteins, peptides, and genetic material, through a TDDS.

Supplementary Materials: The following are available online at http://www.mdpi.com/1999-4923/12/4/392/s1, Figure S1: Effect of [Ch][Ole] content at a fixed S/Co_{mix} concentration (15 wt.%) on the miscibility of [Ch][CA] ILs in a S/Co_{mix}/IPM system at 25 °C; (mean ± SD, n = 3). Figure S2. Phase behavior studies of IL/S/Co_{mix}/IPM MEs consisting of [Ch][Pro] with varying S/Co weight ratios (A) 3:2 (B) 2:3, and (C) 1:1:1 ([Ch][Ole]: Tween-80: Span-20) at 25 °C. Figure S3. Phase behavior studies of IL/S/Co_{mix}/IPM MEs consisting of (A) [Ch][Pro], (B) [Ch][For], and (C) [Ch][Lac] at a 2:1 weight ratio of S/Co at 25 °C. Figure S4. The size and size distribution of IL/S/Co_{mix}/IPM ME (consisting of 15 wt.% S/Co_{mix}, at a 2:1 weight ratio) with different R values (R = molar ratio of IL and S/Co_{mix}) at 25 °C. Figure S5. Dependence of the diameter of IL/S/Co_{mix}/IPM ME (consisting of 15 wt.% S/Co_{mix}, at a 2:1 weight ratio) on R (R = molar ratio of IL and S/Co_{mix}) at 25 °C. Figure S6. The size and size distribution of IL/S/Co_{mix}/IPM ME with different S/Co_{mix} concentrations (wt.%) at a 2:1 weight ratio and R = 0.2 at 25 °C. Figure S7. The size and size distribution of ACV-loaded (2 mg/mL) MEs with varying S/Co weight ratios at 25 °C. Figure S8. ACV encapsulation efficiency of MEs after two months. Table 1S. The density and viscosity of MEs. Table S2. FTIR peak shifts of SC after treatment with different MEs (mean ± SD, n = 3).

Author Contributions: Conceptualization—M.G. and M.R.I.; methodology—M.R.I. and M.R.C.; formal analysis—M.R.I.; investigation—M.R.I.; resources—M.G.; data curation—M.R.I.; writing—original draft preparation—M.R.I.; writing—review and editing—M.R.C., R.W., N.K., M.M. and M.G.; visualization—M.R.I.; supervision—M.G.; project administration—M.G.; funding acquisition—M.G. All authors have read and agreed to the published version of the manuscript.

Funding: This study was funded by the Japan Society for the Promotion of Science (JSPS) (KA-KENHI No. JP20K20440) from the Ministry of Education, Culture, Sports, Science, and Technology of Japan.

Acknowledgments: We thank the Government of Japan (MEXT) for providing the scholarship to carry out higher study and research at Kyushu University. We thank M. Watanabe for providing NMR analysis facilities. We also thank Victoria Muir, PhD, from Edanz Group (www.edanzediting.com/ac) for editing a draft of this manuscript.

Conflicts of Interest: The authors declare no conflict of interest.

References

1. Prausnitz, M.R.; Langer, R. Nihms121685. *Nat. Biotechnol.* **2009**, *26*, 1261–1268. [CrossRef] [PubMed]
2. Amjadi, M.; Sheykhansari, S.; Nelson, B.J.; Sitti, M. Recent Advances in Wearable Transdermal Delivery Systems. *Adv. Mater.* **2018**, *30*, 1–19. [CrossRef] [PubMed]
3. Dahlizar, S.; Futaki, M.; Okada, A.; Yatomi, C.; Todo, H.; Sugibayashi, K. Combined Use of N -Palmitoyl-Glycine-Histidine Gel and Several Penetration Enhancers on the Skin Permeation and Concentration of Metronidazole. *Pharmaceutics* **2018**, *10*, 163. [CrossRef] [PubMed]
4. Karande, P.; Mitragotri, S. Biochimica et Biophysica Acta Enhancement of transdermal drug delivery via synergistic action of chemicals. *Biochim. Biophys. Acta-Biomembr.* **2009**, *1788*, 2362–2373. [CrossRef]
5. Zhang, Y.; Hu, H.; Jing, Q.; Wang, Z.; He, Z.; Wu, T. Improved Biosafety and Transdermal Delivery of Aconitine via Diethylene Glycol Monoethyl Ether-Mediated Microemulsion Assisted with Microneedles. *Pharmaceutics* **2020**, *12*, 163. [CrossRef]
6. Kong, Q.; Higasijima, K.; Wakabayashi, R.; Tahara, Y.; Kitaoka, M.; Obayashi, H.; Hou, Y.; Kamiya, N.; Goto, M. Transcutaneous delivery of immunomodulating pollen extract-galactomannan conjugate by solid-in-oil nanodispersions for pollinosis immunotherapy. *Pharmaceutics* **2019**, *11*, 563. [CrossRef]
7. Xuan, X.Y.; Cheng, Y.L.; Acosta, E. Lecithin-linker microemulsion gelatin gels for extended drug delivery. *Pharmaceutics* **2012**, *4*, 104–129. [CrossRef]
8. Moniruzzaman, M.; Kamiya, N.; Goto, M. Ionic liquid based microemulsion with pharmaceutically accepted components: Formulation and potential applications. *J. Colloid Interface Sci.* **2010**, *352*, 136–142. [CrossRef]
9. Moniruzzaman, M.; Tamura, M.; Tahara, Y.; Kamiya, N.; Goto, M. Ionic liquid-in-oil microemulsion as a potential carrier of sparingly soluble drug: Characterization and cytotoxicity evaluation. *Int. J. Pharm.* **2010**, *400*, 243–250. [CrossRef]
10. Kandasamy, S.; Moniruzzaman, M.; Sivapragasam, M.; Rashid, M.; Ibrahim, M.; Mutalib, A. Separation and Puri fi cation Technology Formulation and characterization of acetate based ionic liquid in oil microemulsion as a carrier for acyclovir and methotrexate. *Sep. Purif. Technol.* **2018**, *196*, 149–156. [CrossRef]
11. Wang, C.; Zhu, J.; Zhang, D.; Yang, Y.; Zheng, L.; Qu, Y.; Yang, X.; Cui, X. Ionic liquid – microemulsions assisting in the transdermal delivery of Dencichine: Preparation, in-vitro and in-vivo evaluations, and investigation of the permeation mechanism. *Int. J. Pharm.* **2018**, *535*, 120–131. [CrossRef] [PubMed]
12. Poh, Y.; Ng, S.; Ho, K. Formulation and characterisation of 1-ethyl-3-methylimidazolium acetate-in-oil microemulsions as the potential vehicle for drug delivery across the skin barrier. *J. Mol. Liq.* **2019**, *273*, 339–345. [CrossRef]
13. Som, I.; Bhatia, K.; Yasir, M. Status of surfactants as penetration enhancers in transdermal drug delivery. *J. Pharm. Bioalied Sci.* **2012**, *4*, 2–9. [CrossRef]
14. Sajid, M.; Sarfaraz, M.; Alam, N.; Raza, M. Preparation, Characterization and Stability Study of Dutasteride Loaded Nanoemulsion for Treatment of Benign Prostatic Hypertrophy. *Iran. J. Pharm. Res.* **2014**, *13*, 1125–1140.
15. Rao, V.G.; Mandal, S.; Ghosh, S.; Banerjee, C.; Sarkar, N. Ionic Liquid-in-Oil Microemulsions Composed of Double Chain Surface Active Ionic Liquid as a Surfactant: Temperature Dependent Solvent and Rotational Relaxation Dynamics of Coumarin-153 in [Py][TF2N]/[C4mim][AOT]/Benzene Microemulsions. *J. Phys. Chem. B* **2012**, *116*, 8210–8221. [CrossRef]
16. Zech, O.; Thomaier, S.; Kolodziejski, A.; Touraud, D.; Grillo, I.; Kunz, W. Ionic Liquids in Microemulsions—A Concept to Extend the Conventional Thermal Stability Range of Micro ACHTUNG RE emulsions. *Chem. Eur. J.* **2010**, *16*, 783–786. [CrossRef]
17. Sidat, Z.; Marimuthu, T.; Kumar, P.; Toit, L.C.; Kondiah, P.P.D.; Choonara, Y.E.; Pillay, V. Ionic Liquids as Potential and Synergistic Permeation Enhancers for Transdermal Drug Delivery. *Pharmaceutics* **2019**, *11*, 96. [CrossRef]
18. Gomes, J.M.; Silva, S.S.; Reis, R.L. Biocompatible ionic liquids: Fundamental behaviours and applications. *Chem. Soc. Rev.* **2019**, *48*, 4317–4335. [CrossRef]
19. Chowdhury, M.R.; Moshikur, R.M.; Wakabayashi, R.; Tahara, Y.; Kamiya, N.; Moniruzzaman, M.; Goto, M. Ionic-Liquid-Based Paclitaxel Preparation: A New Potential Formulation for Cancer Treatment. *Mol. Pharm.* **2018**, *15*, 2484–2488. [CrossRef]

20. Sivapragasam, M.; Moniruzzaman, M.; Goto, M. An Overview on the Toxicological Properties of Ionic Liquids toward Microorganisms. *Biotechnol. J.* **2020**, *1900073*, 1–9. [CrossRef]
21. Santos de Almeida, T.; Júlio, A.; Saraiva, N.; Fernandes, A.S.; Araújo, M.E.M.; Baby, A.R.; Rosado, C.; Mota, J.P. Choline- versus imidazole-based ionic liquids as functional ingredients in topical delivery systems: Cytotoxicity, solubility, and skin permeation studies. *Drug Dev. Ind. Pharm.* **2017**, *43*, 1858–1865. [CrossRef] [PubMed]
22. Gouveia, W.; Jorge, T.F.; Martins, S.; Meireles, M.; Carolino, M.; Cruz, C.; Almeida, T.V.; Araújo, M.E.M. Toxicity of ionic liquids prepared from biomaterials. *Chemosphere* **2014**, *104*, 51–56. [CrossRef] [PubMed]
23. Chowdhury, M.R.; Moshikur, R.M.; Wakabayashi, R.; Tahara, Y.; Kamiya, N.; Moniruzzaman, M.; Goto, M. Development of a novel ionic liquid-curcumin complex to enhance its solubility, stability, and activity. *Chem. Commun.* **2019**, *55*, 7737–7740. [CrossRef] [PubMed]
24. Muhammad, N.; Hossain, M.I.; Man, Z.; El-Harbawi, M.; Bustam, M.A.; Noaman, Y.A.; Mohamed Alitheen, N.B.; Ng, M.K.; Hefter, G.; Yin, C.Y. Synthesis and physical properties of choline carboxylate ionic liquids. *J. Chem. Eng. Data* **2012**, *57*, 2191–2196. [CrossRef]
25. Wu, X.; Chen, Z.; Li, Y.; Yu, Q.; Lu, Y.; Zhu, Q.; Li, Y. Improving dermal delivery of hydrophilic macromolecules by biocompatible ionic liquid based on choline and malic acid. *Int. J. Pharm.* **2019**, *558*, 380–387. [CrossRef]
26. Ali, M.K.; Moshikur, R.M.; Wakabayashi, R.; Tahara, Y.; Moniruzzaman, M.; Kamiya, N.; Goto, M. Synthesis and characterization of choline–fatty-acid-based ionic liquids: A new biocompatible surfactant. *J. Colloid Interface Sci.* **2019**, *551*, 72–80. [CrossRef]
27. Tahara, Y.; Honda, S.; Kamiya, N.; Piao, H.; Hirata, A.; Hayakawa, E. A solid-in-oil nanodispersion for transcutaneous protein delivery. *J. Control. Release* **2008**, *131*, 14–18. [CrossRef]
28. Islam, M.R.; Chowdhury, M.R.; Wakabayashia, R.; Taharaa, Y.; Kamiya, N.; Moniruzzaman, M.; Goto, M. Choline and amino acid based biocompatible ionic liquid mediated transdermal delivery of the sparingly soluble drug acyclovir. *Int. J. Pharm.* **2020**, *582*, 119335. [CrossRef]
29. Zakrewsky, M.; Lovejoy, K.S.; Kern, T.L.; Miller, T.E.; Le, V.; Nagy, A.; Goumas, A.M.; Iyer, R.S.; DelSesto, R.E.; Koppisch, A.T.; et al. Ionic liquids as a class of materials for transdermal delivery and pathogen neutralization. *Proc. Natl. Acad. Sci. USA* **2014**, *111*, 13313–13318. [CrossRef]
30. Chowdhury, M.R.; Moshikur, R.M.; Wakabayashi, R.; Tahara, Y.; Kamiya, N.; Moniruzzaman, M.; Goto, M. In vivo biocompatibility, pharmacokinetics, antitumor efficacy, and hypersensitivity evaluation of ionic liquid-mediated paclitaxel formulations. *Int. J. Pharm.* **2019**, *565*, 219–226. [CrossRef]
31. Petkovic, M.; Ferguson, J.L.; Gunaratne, H.Q.N.; Ferreira, R.; Leit, M.C.; Seddon, K.R.; Rebelo, N.; Silva, C. Novel biocompatible cholinium-based ionic liquids—Toxicity and biodegradability †. *Green Chem.* **2010**, *12*, 643–649. [CrossRef]
32. Gonzalez-Garcia, R.A.; Mccubbin, T.; Navone, L.; Stowers, C.; Nielsen, L.K.; Marcellin, E. Microbial Propionic Acid Production. *Fermentation* **2017**, *3*, 21. [CrossRef]
33. Abe, M.; Kuroda, K.; Sato, D.; Kunimura, H.; Ohno, H. Effects of polarity, hydrophobicity, and density of ionic liquids on cellulose solubility. *Phys. Chem. Chem. Phys.* **2015**, *17*, 32276–32282. [CrossRef] [PubMed]
34. Fukaya, Y.; Sugimoto, A.; Ohno, H. Superior solubility of polysaccharides in low viscosity, polar and halogen-free 1,3-dialkylimidazolium formates. *Biomacromolecules* **2006**, *7*, 3295–3297. [CrossRef] [PubMed]
35. Ali, M.K., Moshikur, R.M.; Wakabayashi, R.; Moniruzzaman, M.; Kamiya, N.; Goto, M. Biocompatible Ionic Liquid Surfactant-Based Microemulsion as a Potential Carrier for Sparingly Soluble Drugs. *ACS Sustain. Chem. Eng.* **2020**. [CrossRef]
36. Wang, W.; Wei, H.; Du, Z.; Tai, X.; Wang, G. Formation and Characterization of Fully Dilutable Microemulsion with Fatty Acid Methyl Esters as Oil Phase. *ACS Sustain. Chem. Eng.* **2015**, *3*, 443–450. [CrossRef]
37. Pramanik, R.; Ghatak, C.; Rao, V.G.; Sarkar, S.; Sarkar, N. Room Temperature Ionic Liquid in Confined Media: A Temperature Dependence Solvation Study in [bmim][BF_4]/BHDC/Benzene Reverse Micelles. *J. Phys. Chem. B* **2011**, *115*, 5971–5979. [CrossRef]
38. Cavalcanti, A.L.M.; Reis, M.Y.F.A.; Silva, G.C.L.; Ramalho, Í.M.M.; Guimarães, G.P.; Silva, J.A.; Saraiva, K.L.A.; Damasceno, B.P.G.L. Microemulsion for topical application of pentoxifylline: In vitro release and in vivo evaluation. *Int. J. Pharm.* **2016**, *506*, 351–360. [CrossRef]
39. Lu, D.; Rhodes, D.G. Mixed composition films of Spans and Tween 80 at the air-water interface. *Langmuir* **2000**, *16*, 8107–8112. [CrossRef]

40. Kantarci, G.; Özgüney, I.; Karasulu, H.Y.; Arzik, S.; Güneri, T. Comparison of different water/oil microemulsions containing diclofenac sodium: Preparation, characterization, release rate, and skin irritation studies. *AAPS PharmSciTech* **2007**, *8*, 1–7. [CrossRef]
41. Narang, A.S.; Delmarre, D.; Gao, D. Stable drug encapsulation in micelles and microemulsions. *Int. J. Pharm.* **2007**, *345*, 9–25. [CrossRef] [PubMed]
42. Chen, H.; Chang, X.; Weng, T.; Zhao, X.; Gao, Z. A study of microemulsion systems for transdermal delivery of triptolide. *J. Control. Release* **2004**, *98*, 427–436. [CrossRef] [PubMed]
43. Huang, Y.R.; Lin, Y.H.; Lu, T.M.; Wang, R.J.; Tsai, Y.H.; Wu, P.C. Transdermal delivery of capsaicin derivative-sodium nonivamide acetate using microemulsions as vehicles. *Int. J. Pharm.* **2008**, *349*, 206–211. [CrossRef] [PubMed]
44. Panchagnula, R.; Desu, H.; Jain, A.; Khandavilli, S. Feasibility studies of dermal delivery of paclitaxel with binary combinations of ethanol and isopropyl myristate: Role of solubility, partitioning and lipid bilayer perturbation. *Farmaco* **2005**, *60*, 894–899. [CrossRef] [PubMed]
45. Sintov, A.C.; Shapiro, L. New microemulsion vehicle facilitates percutaneous penetration in vitro and cutaneous drug bioavailability in vivo. *J. Control. Release* **2004**, *95*, 173–183. [CrossRef] [PubMed]
46. Schwarz, J.C.; Pagitsch, E.; Valenta, C. Comparison of ATR-FTIR spectra of porcine vaginal and buccal mucosa with ear skin and penetration analysis of drug and vehicle components into pig ear. *Eur. J. Pharm. Sci.* **2013**, *50*, 595–600. [CrossRef]
47. Zhang, Y.; Chen, X.; Li, X.; Zhong, D. Development of a liquid chromatographic-tandem mass spectrometric method with precolumn derivatization for the determination of dencichine in rat plasma. *Anal. Chim. Acta* **2006**, *566*, 200–206. [CrossRef]
48. Banerjee, A.; Ibsen, K.; Iwao, Y.; Zakrewsky, M.; Mitragotri, S. Transdermal Protein Delivery Using Choline and Geranate (CAG) Deep Eutectic Solvent. *Adv. Healthc. Mater.* **2017**, *6*, 1–11. [CrossRef]

Review

Ionic Liquids as Potential and Synergistic Permeation Enhancers for Transdermal Drug Delivery

Zainul Sidat, Thashree Marimuthu, Pradeep Kumar, Lisa C. du Toit, Pierre P. D. Kondiah, Yahya E. Choonara and Viness Pillay *

Wits Advanced Drug Delivery Platform Research Unit, Department of Pharmacy and Pharmacology, School of Therapeutic Sciences, Faculty of Health Sciences, University of the Witwatersrand, Johannesburg, 7 York Road, Parktown 2193, South Africa; zainul.sidat@students.wits.ac.za (Z.S.); thashree.marimuthu@wits.ac.za (T.M.); pradeep.kumar@wits.ac.za (P.K.); lisa.dutoit1@wits.ac.za (L.C.d.T.); pierre.kondiah@wits.ac.za (P.P.D.K.); yahya.choonara@wits.ac.za (Y.E.C.)

* Correspondence: viness.pillay@wits.ac.za; Tel.: +27-11-717-2274

Received: 25 December 2018; Accepted: 15 February 2019; Published: 22 February 2019

Abstract: Transdermal drug delivery systems (TDDS) show clear advantages over conventional routes of drug administration. Nonetheless, there are limitations to current TDDS which warrant further research to improve current TDD platforms. Spurred by the synthesis of novel biodegradable ionic liquids (ILs) and favorable cytotoxicity studies, ILs were shown to be a possible solution to overcome these challenges. Their favorable application in overcoming challenges ranging from synthesis, manufacture, and even therapeutic benefits were documented. In this review, said ILs are highlighted and their role in TDDS is reviewed in terms of (a) ILs as permeation enhancers (single agents or combined), (b) ILs in drug modification, and (c) ILs as active pharmaceutical ingredients. Furthermore, future combination of ILs with other chemical permeation enhancers (CPEs) is proposed and discussed.

Keywords: ionic liquids; transdermal; synergy; permeation enhancer; chemical; physical; Transdermal drug delivery systems

1. Introduction

Currently, ionic liquids (ILs) are a class of compounds under intensive investigation for a multitude of applications including, but not limited to, green chemistry, chemical synthesis [1], catalysis, lubricant fluids [2], plasticizers, organic solvent replacement [3], electrochemistry, and bio- and nano-technologies, among many more [1–6]. However, of particular importance is the application of ILs for biomedical applications [4]—more specifically transdermal drug delivery [4]. Based on the reported use of ILs as chemical permeation enhancers (CPEs), there is continued interest for ILs in transdermal drug delivery. ILs were shown to enhance transdermal transcellular and paracellular transport, bypassing the barrier properties of the stratum corneum (SC), employing mechanisms such as disruption of cellular integrity, fluidization, and creation of diffusional pathways and extraction of lipid components in the SC [7,8].

The ability for room-temperature ILs (RTIL) to be modified for various purposes allows for them to be used in many settings, such as solubilizing agents with the ability to solubilize a wide variety of compounds [4], or as permeation enhancers which act on biological membranes leading to improved efficacy and clinical outcomes [9–14]. Early work done by Moniruzzaman et al. [15] (2010) showed that IL soluble drug acyclovir can form stable IL–oil microemulsions with dimethylimidazolium dimethylphosphate ($MMIM^+$)($MMPO_4^-$), allowing for an alternative solvent system to be applied when solubilizing drugs with low solubility in organic solvents. Building on this work, Moniruzzaman and co-workers (2016) also applied the solubilizing properties of ILs to biomolecules such as proteins, enzymes, and DNA [16]. Moreover, etodolac, a poorly water-soluble drug, was formulated with 1-butyl-3-methylimidazolium hexafluorophosphate ($BMIMPF_6$) to give an ionic-liquid-in-water (IL/w) microemulsion (ME) [17].

The solvating power of ILs is remarkable, leading to their use not only in topical systems, but also oral systems, and their applications in drug delivery formulations are not only limited to solubilizing agents. They were also used to overcome synthesis challenges by inhibiting unwanted polymorphs in crystalline active pharmaceutical ingredients (APIs). Enhanced therapeutic outcomes were noted when strategies such as formulating APIs as ILs (API-IL) or the use of pharmacologically active ILs were applied, including, but not limited to, anti-bacterial, -viral, and -fungal activity, cytotoxic agents, and biofilm-disrupting agents.

This review provides an account of the successful implementation and application of ILs in current transdermal systems and how they can be leveraged for enhanced outcomes, along with their applications in all phases of drug delivery including fabrication of transdermal systems, as an excipient to enhance formulations providing permeation, and even as therapeutic moieties to improve disease outcomes. This review is based on mechanistic, computational, and structure–activity evidence for their use in improving permeation and strategies for synergistic combinational therapy to that end.

A Brief Note on Deep Eutectic Solvents (DES)

Both DESs and ILs are touted to be greener solvents than current industrial standards; they are similar in nature with the key difference that can be found in the components used. In DESs, an organic halide is complexed with an organic agent, most often a hydrogen-bond donor derived from non-ionic species [18,19]. Similarities between both are significant, including their applications as modifiable solvents where DESs are gaining more and more traction over ILs due to a host of factors. The most notable reasons include decreased cost, relatively easy preparation, lack of water reactivity (a major disadvantage of some of the earlier ILs), non-toxicity, and biodegradation pathways (a major advantage as several ILs show environmental damage) [18–20]. The hydrogen bonding seemingly confers many advantages similar to ILs, providing compounds that have low vapor pressure, dipolar nature, low volatility, and more [18]. While they are currently being applied to transdermal delivery systems, DESs are beyond the scope of this review and are, therefore, omitted.

2. Properties of ILs

ILs are an interesting group of chemical compounds composed of ions which have melting points below 100 °C [21–23]. ILs are generally described as having ideal properties, such as low to negligible vapor pressure at room temperature, extensive and varying solubility profiles, high thermal stability, non-flammability, adaptable polarity, inert chemical profiles, exceptionally low melting points (for ionic-bonded compounds), variable viscosities, and many more customizable properties [14,21,23,24]. Pioneering work by Paul Walden in 1914 established ethyl-ammonium nitrate as a prototype IL which was later succeeded by a multitude of alternative generations of ILs over several decades [1] with the emergence of new tailored and greener ILs [21].

While ILs are often discussed as a group and are attributed relative group properties, it is essential to note that the sheer diversity of ILs [23] means that there are very few descriptors that fit all ILs [22]. They can be modified to provide a variety of possible properties that are desired. They are made up of two distinct components—the anion and cation moieties. The cation is generally bulky and organic in nature, whereas the anion is relatively much smaller and inorganic. This combination leads to a decrease in the crystallinity of the system, allowing ILs to be liquid at such low temperatures and confer any number of favorable properties as mentioned above.

The make-up of ILs means that any number of permutations are possible so long as a cationic and anionic species with poor crystalline packing exist. This led to the discovery of many nascent species. Included among these newer IL bases (Figure 1) are cholinium and guanidinium cations, which confer the advantage of being biodegradable and are, therefore, referred to as bioinspired ILs (BILs) [25]. The more traditional base cations include quaternary ammonium, imidazolium, pyrrolidinium, pyridinium, and phosphonium cations. Some of the newer species, such as metal-containing cations, no longer fit in the traditional scope of ILs where organic cations and inorganic anions are combined. These new metal ILs are formulated using metal-containing cations and various anions, allowing the formation of IL-like salts. A further subset of these metal-containing ILs involves those with magnetic properties [26]. The abovementioned cation species and new alternatives increase the scope for application of these ILs in biomedical research—particularly bioinspired IL bases.

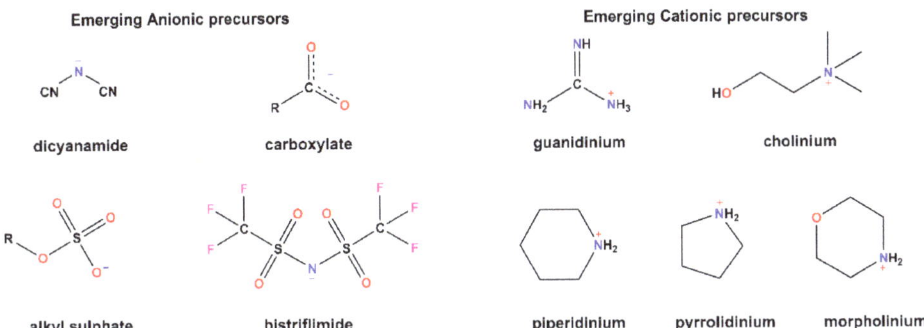

Figure 1. Emerging anions and cations used in ionic liquids (ILs).

2.1. Chemical Traits

Due to the sheer number of possible permutations of these constituents that make up ILs, it becomes very difficult to investigate possible chemical effects [14]. However multiple resources including, but not limited to, computational models, databases such as IL-THERMO, and studies of effects (using theoretical methods. i.e., classical and density functional theory (DFT) molecular dynamics modeling), gave insights into chemical properties.

Recent studies involving the structural properties of ILs showed that ILs contain micro- and nano-domain organizational structures [27], and ions tend to self-assemble into subdomains forming amphiphilic nano-structures which persist. Based on these structural features, ILs seem to resemble more of a nano-heterogenous substance which is a largely unique characteristic of ILs and is not found commonly among solvents and other dissolution media. Spatial studies on these domains are extremely difficult and are modeled theoretically using classical and DFT molecular dynamics. The work is confirmed by spectral studies and accounts for difficulties when certain ILs are characterized by nuclear magnetic resonance spectroscopy (NMR), thus requiring external forces throughout the characterization process disrupting ultra-structures [27].

Hydrogen bonding and ionic interactions are the key drivers in structural organization found in ILs. A core difference between ions of inorganic salts and ILs is the size of the participant ions. Inorganic salts consist of spherical species allowing neat and ordered packing, in contrast to the large and bulky cation of ILs with the charge being distributed over a much larger area. This in turn influences the local structure of the IL, and minute changes can lead to drastic changes in the activity demonstrated readily by cation choice. As an essential component, the cation influences factors such as viscosity, conductance, electrostatic forces, surfactant properties, polarity, aggregation challenges, and toxicity profiles, among others. A simple variation such as chain length can have extreme influences on the physio-chemical properties of the resulting ILs (as outlined in Table 1).

The ultra-level organization of ILs is only possible due to the large bulky cation. The charge distribution over a much larger area allows for lower charge density and thwarts electrostatic repulsion. This convergence creates noncovalent interactions, leading to the structural properties observed in ILs.

The application of ILs as solvents is a promising trend in current research, and the solubility of ILs is a key concern. Solvent–solute interactions for ILs are often specific rather than explained in non-specific class terms. The ability for ILs to be miscible with a wide variety of other solvents provides an ideal solution for dissolving most solutes. The general mechanism for IL solvation is not yet clearly understood, as some undergo H-bonding, others show dipole–dipole interactions, and some even undergo π–π interactions. An interesting find by Visser et al. [28] (2002) showed that neutral substances dissolve with ease as opposed to ionic species; however, with optimization, they are capable of dissolving most substances to give ideal miscibility in reasonable molar ratios [29].

Table 1. Singular change to cation chain length leading to altered physicochemical profile. IL—ionic liquid.

Chain Length Alteration	Change on Physio-Chemical Properties of ILs	Reference
Increase	Increased viscosity	[6,30]
	Increase enthalpy of vaporization	[6]
	Increased aggregation (not necessarily ordered)	[6,30,31]
	Increased toxicity (bacterial and marine ecosystems)	[3,32]
	Increased surfactant activity	[6]
Chain lengths similar to biological membranes	Increased bioaccumulation (potential for toxicity)	[3,20]
Decrease	Increased conductance	[6]
	Increase in electrostatic forces	[6,30]
	Ordered aggregation; depends largely on polarity and geometric packing	[6,31]
	Increased lipase catalytic activity	[19,33]
	Increased polarity	[6]

2.2. Physical Properties

Physical properties of ILs are as diverse as chemical properties, and most of these characteristics can be configured and fine-tuned for a given application [12]. Among the most important properties is the density of ILs. The density of ILs ranges between 0.9 and 1.7 $g \cdot cm^{-3}$ which renders ILs to be relatively more dense than conventional organic solvents [34]. Density can be affected by temperature, pressure, and, most importantly, molecular mass and the interactions between molecules. As such, it was found that larger organic cation-containing ILs have a lower density [29].

Another important property affecting the application of ILs in TDD is viscosity. The viscosity of a given IL is a very important property, which varies from <10 to >1000 cP at room temperature. This characteristic of ILs limits the possible applications of ILs where they are used as solvents. Viscosity for ILs is expressed as a viscosity co-efficient rather than a stated value; this in large part is due to ILs not following Arrhenius behavior. The alternative, Vogel–Tammann–Fulcher (VTF) equation, accounts for an additional factor, i.e., the glass transition.

The lower melting points of ILs despite their inherent ionic nature are among the more fascinating properties of ILs. Due to this property and low vapor pressures, they are extremely attractive alternatives to current organic solvents. RTILs are defined by melting points at or below room temperatures, and variances are accounted for due to charge distribution, H-bonding capability, and symmetry. Freezing-point calculations are often experimental and unreliable as ILs undergo variable rates of supercooling. Factors affecting these properties were experimentally found by Katritzky et al. [35] (2002), and include molecular shape, symmetry, rotational freedom, and electrostatic interactions.

Polarity features a key role in solvent–solvate interactions. As explained for each of the properties above and for ILs as a whole, there are no unifying rules; rather, there are structure-related changes that can be tuned and optimized for a given task. Solvatochromic dyes gained favor in polarity studies where a variety of compounds such as Nile-red and betaine-30 have different absorption and emission bands based on the solvents in which they are dissolved. Key features affecting the polarity of ILs include chain length variation, whereby long and branching chains are more hydrophobic.

2.3. Interactions of ILs with Biomolecules and Membranes

The organic character and increased use of ILs in industrial processes inspired the first studies of ILs on bio-molecules and bio-organisms. While several studies highlighted toxicity and their negative impacts, these can be applied in a variety of ways to be advantageous. The possibilities include, but are not limited to, anti-bacterial properties, and the stabilization and storage of bio-molecules such as proteins, enzymes, and DNA. More impressively, the abilities of ILs to re-functionalize defective amyloid fibers, dissolve complex polysaccharides, and create pores in bio-membranes, among many others were reported [4,32,36–40]. These processes can yield merits such as selective bacterial killing, selective toxicity to cancerous cells, and application for transdermal systems, respectively.

Any initial reaction will occur at the interface between a bio-membrane and the ILs. It is critical that the interaction must be safe for human use. These membranes serve not only as a protective barrier, but are also key in regulating diffusional processes and cell replication. The most accepted model of bio-membranes is the phospholipid bilayer, and testing employing ILs investigating their effects employs this model. Several IL bases show significant resemblance to phospholipids and can be synthesized to resemble biologically based materials, as demonstrated by Wang et al. [41] using dimethylimidazolium iodide derivatives (Figure 2). In this study, the similarities extend to include an ionic polar character and internal hydrophobic regions. Increasingly, ILs are being based on biological materials such as those reported in the previous study and other work done by the same authors [42]. Bio-inspired ILs are different to biomimicry ILs such as those reported by these previous studies. Bioinspired ILs such as those reported by Benedetto and workers (2014) [43], where choline-based cations and several amino-acid–anion pairs were modeled and evaluated for biological systems as starting materials. This led to significant similarities between endogenous groups and these artificial substrates that can be observed.

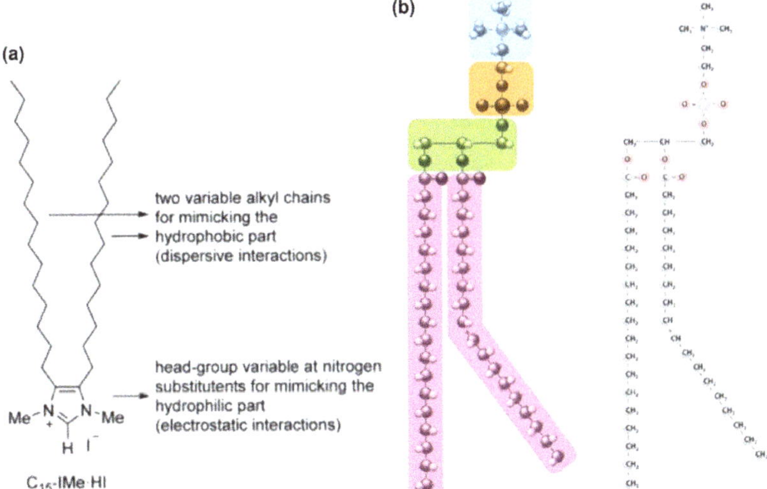

Figure 2. Sketches depicting (**a**) double-tail lipid-mimic imidazolium-based IL; (**b**) glycerol–phospholipid sub-regions with large tail lengths. The similarities are striking and allow for their ability to intercalate within the phospholipid membrane structure. Adapted with permissions from (**a**) Reference [41] John Wiley and Sons, ©2015, and (**b**) https://www.nature.com/scitable/topicpage/cell-membranes-14052567 Nature Education ©2010.

Simple biomimicry between IL structure and endogenous phospholipids, however well done, does not assure safety and compatibility of ILs in biomedical applications. This is evident by the work of Evans et al., where it was reported that slight changes in the cation chain when using 1-butyl-3-methylimidazolium chloride led to substantial damage to the bio-membranes that were investigated [44,45]. These experiments were limited in their scope, and work done by Benedetto et al. [46] further explored the penetration of 1-butyl-3-methyl-imidazolium chloride (BMIMCl) and choline chloride $(Chol^+)(Cl^-)$ ILs in bio-membranes by means of neutron scattering. This work demonstrated that phospholipid bilayers retain their two-dimensional (2D) characteristic configurations at concentrations up to 0.5 M. This work gave significant insight into changes to the bio-membranes which included the shrinking thickness of the bilayer, the accumulation of IL cations within the junction between the polar heads and hydrocarbon tails, and the lipid bilayer composition affecting the amount of accumulation that occurs, as well as the degree of penetration, but not the location of accumulation. Further studies such as crucial work done by Jing et al. [45] also confirmed this bioaccumulation of ILs causing swelling and also showed that concentration was a key concern and could completely disrupt the bilayer. Building on their previous work, Benedetto et al. [47] included molecular dynamics simulations which verified the validity of the results, and a potential mechanism of interaction of ILS with bio-membranes (Figure 3) was proposed.

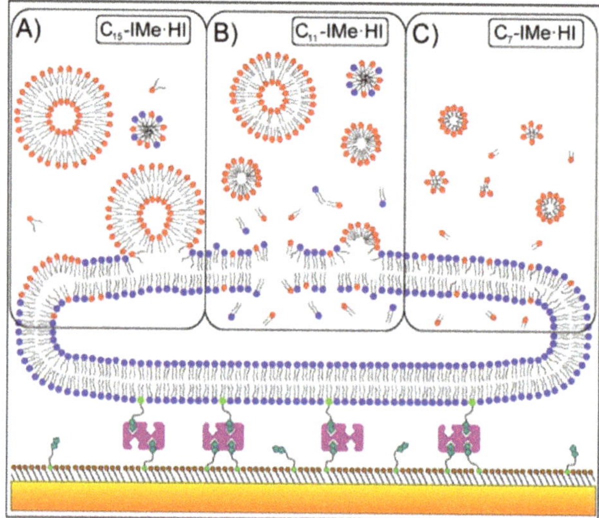

Figure 3. A model of the interaction between ILs (imidazolium-based) and bio-membranes where a gold-coated sensor surface (orange) has a chemisorbed self-assembled monolayer (brown) tethered by biotin linkers (green) to streptavidin which also tethers liposomes (blue). Panel (**A**) depicts a vesicle, which can then associate and intercalate into the bilayer membrane. Panel (**B**) focuses largely on mechanisms causing lyses to the membrane by forming vesicles or micelles in an aqueous medium, thereafter disrupting the membrane structure. Panel (**C**) focuses on mechanisms that allow pass-through within the membrane by disassociating into single molecules. Adapted with permission from Drucker 2017 [48]; © 2017, American Chemical Society.

Effects of ILs on bio-membranes are affected by several factors. The concentration of ILs applied affects the bio-membranes and results in shrinkage of the bilayer and increased the elasticity of the bilayer. Disruption of phospholipid membranes is affected by the hydrocarbon chain length in the cation. The interaction of cholesterol-containing bio-membranes makes them more resistant to rupture, but at a threshold concentration; once exceeded, the effects on these cholesterol-containing bio-membranes are much more apparent. Concentration and chain length play critical roles in cytotoxicity, as they allow easier penetration into the membrane. Short-tailed ILs spontaneously insert into the membrane, but long-tailed ILs self-assembled into micelles that are eventually absorbed and form a monolayer in the upper portion of the bilayer. This means that short-tailed ILs are more mobile and can diffuse more easily than their long-tailed counterparts.

3. Current Challenges in Transdermal Drug Delivery

Transdermal drug delivery is yet to be fully appreciated, as more studies are required to fully realize its full potential [49]. The major challenge associated with transdermal drug delivery is the almost impermeable barrier created by the SC measuring 10 µm in thickness with variances in different parts of the body [50–53]. While steady work is being carried out in this field, there are still very few APIs that are likely candidates for delivery through this route. Permeation of APIs is historically the biggest challenge in transdermal drug delivery, with multiple generations of transdermal delivery systems [49] trying to overcome this without much progress to show [54,55].

Transdermal or percutaneous absorption refers to the rate of absorption of a topically applied chemical. The necessity for determination of this rate is twofold the effectiveness of transdermal application, thereby avoiding many of the disadvantages associated with other modes of drug delivery such as oral or parenteral routes. With so many factors (Figure 4) affecting transdermal drug delivery such as the method and location of drug delivery, the drug molecule itself, and factors such as inter-individual variances (skin age, condition, and blood flow), it becomes apparent that simple zero- or first-order kinetics are not sufficient to describe it. A key consideration in effective drug delivery systems is protecting the drug moiety from undesirable metabolism while enhancing transport to active sites. Biological and physiological hurdles prevent efficient drug delivery, particularly the skin in transdermal systems.

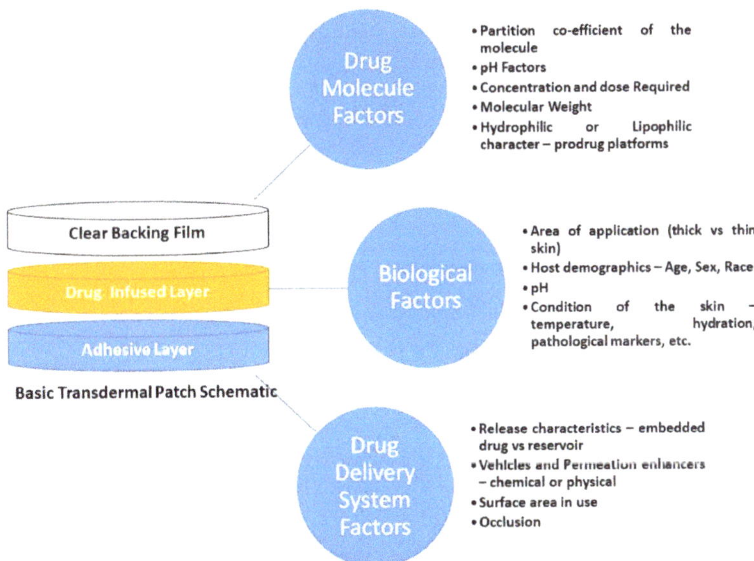

Figure 4. Depiction of the most critical factors to consider when designing a transdermal drug delivery system.

Skin is the primary barrier to any transdermal drug delivery. Designed to be impermeable and extremely adapted to its function [56], the outermost layer (the SC) is the most impermeable layer and extremely well differentiated, derived from keratinocytes that are terminally differentiated. A lipophilic matrix that anchors corneocytes made up of free fatty acids, cholesterol, and ceramides provides the only diffusion phase to allow drug delivery to occur into the subcutaneous layers [57,58]. Freeze-fracture electron microscopy showed that the SC is arranged in lamellae, which is facilitated by the presence of cholesterol sulfate. The matrix is extremely heterogeneous, made up of 11 classes and 342 individual ceramide species alone which were identified by Masukawa and co-workers (2008) [59].

Flux across the skin for molecules that are natively permeable can be simply expressed as

$$J = K_p \cdot \Delta C, \tag{1}$$

where J is the flux (expressed usually as μg·cm^{-2}·h^{-1}), K is a co-efficient denoting the permeability of the skin with regard to the permeant in question, and ΔC the concentration gradient across the barrier. Where passive diffusion is the primary means of permeation, K_p is defined by the partition co-efficient (P) and the diffusion co-efficient (D) and the thickness of the boundary (h), mathematically expressed as

$$K_p = \frac{P \cdot D}{h}. \tag{2}$$

To date, many techniques and methods were evaluated for improving permeation, and none of them can be labeled as ideal. The criteria [60,61] to be labeled as ideal is quite significant and include a varying number of properties. These properties range from chemical nature, such as non-toxic, non-irritant, and non-allergenic, to ideal conditions such as rapid activity and inert physiological profile. This list also includes mechanistic criteria, such as unidirectional biased permeation, i.e., allowing drug molecules in but not letting that which is in the body out, and rapid recovery of skin when the agent is removed. Formulation application is also considered, which must meet standards such as compatibility with a variety of formulations and being cosmetically acceptable for patient compliance.

The criteria for drug molecules suitable for transdermal drug delivery are extremely stringent and include, but are not limited to, an aqueous solubility greater than 1 mg/mL, an oil–water partition co-efficient between 10 and 1000 (at the same time, it cannot be too lipophilic as then it will remain for an extended period in the subcutaneous layer), molecular weight under 500 Da allowing it to be sufficiently mobile, and a melting point under 200 °C with a dose not exceeding 10 mg, i.e., potent drug molecules [57,62,63]. These are preliminary ideal conditions for passive drug delivery across the skin barrier. One of the most widely used indicators is the partition co-efficient described by Potts and Guy [64] denoting hydrophilicity, with optimal results being seen at values ranging from (1)–(3).

$$\log K_p = 0.71 \cdot \log P_{Octanol-water} - 0.0061 \cdot Molecular\ Mass - 2.74. \tag{3}$$

With only a few drug molecules such as scopolamine, nitroglycerin, clonidine, estradiol, fentanyl, nicotine, testosterone, and norethisterone being able to meet the stringent criteria necessary for TDD, it becomes very apparent that the limitations to this system are vast but not without their own advantages. The skin provides a vast surface area for absorption of ~1–2 m^2, decreased metabolism of drugs, enhanced patient compliance due to several factors (non-invasive, reduced side-effect profiles [65]), painless, and extended drug release systems especially for short half-life drug molecules decreasing patient interventions, among others [14,57,63,66]. Those advantages are patient-driven and quite substantial; however, these systems also decrease the need for skilled healthcare practitioners especially in rural settings, and produce limited hazardous waste materials [67].

Three major techniques (with a few pertinent examples listed in Table 2) are used to improve permeation across the skin. The first is formulation enhancement, whereby a variety of formulation modifications and strategies are used to help the delivery system achieve permeation. These strategies are not always applicable and can only be used for those drug molecules which are natively permeable through the skin. The second is physical permeation enhancement. This usually involves usually devices of some sort that disrupt the skin barrier by physical means by producing pores which drug molecules can permeate through. These systems have the disadvantage of needing a device which the patient must carry around; alternatively, the device may cause punctures or cavitation, which carries a risk of infection [68].

The last and most investigated methods are CPEs. This method involves the use of chemical compounds that interact with the skin and affect the activity of the barrier. Various functional groups in these chemicals interact with the skin and its lipid content causing disfiguration to the skin layer. Chemical enhancers of this nature are liable to cause skin irritation, as permeation activity is proportional to irritant activity.

Table 2. Permeation techniques (including formulation enhancement strategies, physical permeation techniques, and chemical permeation enhancers) employed and examples of well-known applications of these techniques.

Permeation Techniques	Example of Technique	References
Formulation enhancement	1) Supersaturated systems 2) Microemulsions 3) Drug moiety modification	[69–71]
Physical permeation techniques	1) Electroporation 2) Sonophoresis 3) Microneedles	[72–75]
Chemical permeation enhancers	1) Alcohols a. Long-chain fatty alcohols b. Short-chain alcohols 2) Fatty acids 3) Sulfoxides 4) Terpenes	[57,76,77]

Current Limitations of Chemical Permeation Enhancers (CPEs)

The most extensively used method of permeation enhancement involves the use of chemical moieties that would interfere with functional groups in the skin in a reversible manner, thereby compromising barrier activity in the SC. This method can be applied to permeate otherwise impermeable drug molecules and to significantly enhance those that already can permeate through. When considering chemical permeation enhancement, several factors play a key role. The factors can be narrowed to the steady-state flux $\left(\frac{dm}{dt}\right)$ measured by the mass (m) passing per unit area of membrane in time (t), concentration of permeant (C_0), partition co-efficient (K) between the membrane and application, diffusion co-efficient (D), and membrane thickness (h), expressed as [78]

$$\frac{dm}{dt} = \frac{D \cdot C_0 \cdot K}{h}. \tag{4}$$

The advantages of using CPEs over other types of enhancers include factors such as design flexibility, self-administration, patient compliance, and easy incorporation into formulations [79]. The inherent limitations to CPEs are quite significant, especially when considering that high-molecular-weight compounds are precluded. Many recent advances are based on peptides and protein deliveries which are, by their nature, large in size and often charged species. This list does not preclude the use of multiple agents (including physical permeation enhancers) to act synergistically [54]. Many are used in combination to potentiate the effects of each other such as the combinations of ethanol and propylene–glycol. However, each of these chemicals has the potential to be irritant to the skin [63]; therefore, the use of some of these agents and any combination increases the risk of skin irritation that may be deemed clinically unacceptable. Chemical enhancers that are found in marketed dermatological products are usually alcohols (ethanol), propylene glycol, and sodium lauryl sulfate [57].

4. ILs Meeting the Needs in Transdermal Drug Delivery

4.1. ILs as Skin Permeation Enhancers

ILs were shown to enhance permeation of drugs through the skin [15,40,80–83]; therefore, many research studies were conducted to account for their underlying mechanisms of action. Several mechanisms that are largely dependent on the chemical make-up of the IL were proposed. Most notably, the work done by Monti and colleagues showed structure to have a large impact on the degree of permeation [84]. A key factor related to permeation enhancement is the physicochemical properties of ILs. Recent research suggests electronic profiles of the ILs have a large part to play in their permeation activity [85]. However, this mechanism is extremely broad and does not account for all IL permeation enhancement profiles. Broadly classified, all ILs with permeation enhancement are hydrophilic or hydrophobic. Hydrophilic ILs act by opening tight junctions within the SC, thereby promoting paracellular transport (Figure 5) acting by enhancing fluidization mainly within protein and lipid regions, whereas hydrophobic ILs improve partitioning into the epithelial membrane by providing channels, thus promoting transcellular transport in the lipid regions [25]. Among the best documented is the activity of 1-octyl-3-methylimidazolium-based ILs which act by disrupting structural integrity by inserting into the membrane [7]. It was also demonstrated that ILs possess the ability to fluidize cell membranes, particularly seen with hydrophilic imidazolium-based ILs [8,40], as well as lipid extraction in the SC. Transdermal delivery of several unlikely drug candidates such as protein molecules [86,87], methotrexate [88], and acyclovir [89], among others benefited greatly from IL incorporation, opening the way for a multitude of alternative possible drug molecules.

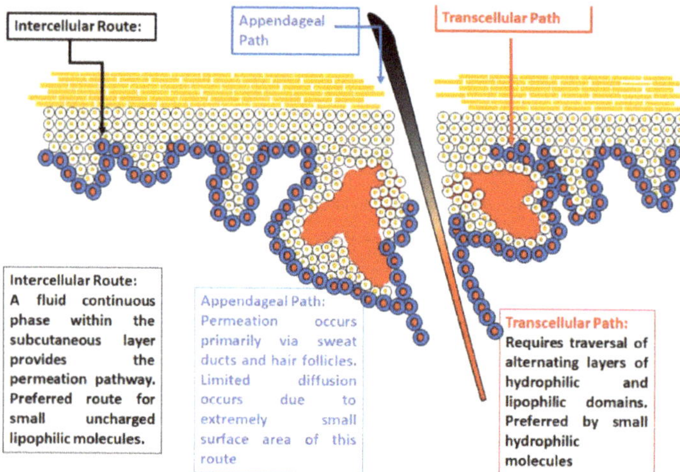

Figure 5. Depiction of the possible routes of transdermal permeation which lead to effective transdermal absorption.

Toxicity is a crucial factor to consider when looking at the medical or biomedical applications of any technology. Older generations of ILs were not suitable for medical applications. To overcome this drawback, bioinspired ILs were designed and were shown to give desirable biodegradation and decreased toxicity profiles. Choline-based bioinspired ILs are currently the most intensively investigated [40,90]. Cytotoxicity studies conducted, usually focusing on ecological toxicity, do indeed provide a reason to be concerned, especially amidst claims that some ILs are more toxic than current standards [91]. Toxicity correlation exists, whereby longer chain lengths are found to be more toxic, as well as effects due to the presence of pharmacological activity, the present ionic species (anion and cations), and the presence of oxygen in the compound. Limited studies were performed in an effort to gauge toxicity on humans, and they include enzymatic assays (acetylcholinesterase (AchE) and AMP deaminase), as well as cytotoxicity studies against cancerous cell lines (colonic, cervical, and breast) [92] and some primary human cells such as normal human bronchial epithelial lines [40].

However, these abovementioned studies show toxicity on single-cell organisms, which can be exploited for antimicrobial properties [40]. Additionally, the use of ILs as adjunctive treatments or alternatives to chemotherapy on cancer cell lines is possible and was reported in Reference [4]. Cholinium-based ILs were proven to be much less toxic than regular ILs, whereby effects on AchE are much more limited, and persistence in the environment is limited by the rapid and substantial biodegradation [14,93]. These cholinium bioinspired ILs have limited or no applications in antimicrobials or antineoplastic agents.

The mechanism through which ILs can be beneficial to transdermal systems is not only limited to their permeation enhancement. They can play roles such as surfactants and optimize the thermodynamic activity of drug molecules, they can act as efficient solubilizing agents or cause fluidization of the lipid bilayer, and they can even disrupt the matrix by acting on keratin fibrils. The permeation activity is high with limited cytotoxicity, which fulfils two of the criteria for ideal permeation enhancers [84].

4.2. Evidence of Successful Synergy between Combinations of ILs and Chemical Enhancers

Synergistic combinations of ILs with other chemical enhancers are limited and mostly experimental in nature. The majority of IL applications do not lie in medical and biomedical use; thus, sparse work was done beyond proving IL effectiveness as a penetration enhancer, and limited toxicity studies make the investigation of ILs in medical applications seem preliminary [94]. Most combinational studies look at ILs, such as use as solvents, reaction media, and excipient substitution in final formulations.

The versatility of ILs does not come from their properties, but rather their heterogeneous nature, allowing them to be synthesized from any number of sources. So long as the result is an IL, it will retain many of the advantages of ILs. They can be synthesized from dual APIs, from biologically based ions, and even from normal synthetic compounds and solvents. This means that the proven safety of GRAS (generally recognized as safe) chemicals (such as terpenes and surfactants) can be leveraged as a basis for a safe and effective ILs. This is not to say that ILs are not safe; in fact, one of the major selling points of ILs is safety; however, limited safety data make them risky investments in the current market for commercial applications.

The current market CPEs can benefit from a variety of ways when incorporated with ILs (as explored briefly in Table 3. The general properties can be improved, as well as enhancing permeation, reduction of unwanted side-effects, and better safety profiles. Successful synergistic combinations can be proposed ad nauseum due to the versatile nature of ILs; however, in this paper we highlight some of the more practical possibilities which are based on previous or extrapolated evidence for their inclusion in pharmaceutical products for higher standards.

IL incorporation with alcohol-based permeation enhancers can lead to the incorporation of a wider variety of APIs via co-solvency, which aids in drug solubility. Similarly, with other permeation enhancers such as Azone or terpenes, limitations when incorporating APIs can be overcome with IL combination. Fatty-acid permeation is potent after a pre-treatment cycle; by using this method and via the incorporation of APIs in IL vehicles, not only can the permeation be enhanced, but the rate of transport can also be increased. Dimethyl sulfoxide (DMSO) incorporation into ILs may overcome limitations such as lowered concentration of DMSO for permeation activity and decreased side-effect profiles; however, the tangible presence of DMSO reduces the discreet profile of the IL and, thus, should be considered in sealed systems. Surfactants are versatile and relatively safe and are, therefore, reused in many formulations. They also boast extremely high permeation enhancement effects; however, the most active permeation surfactants are also extremely irritant to the skin as demonstrated with cationic and anionic surfactants. In this case, finding a balance between irritant properties and activity is achieved with non-ionic surfactants. The incorporation of safer non-ionic surfactants into ILs enhances permeant activity while maintaining safety profiles. Some of the earliest work with medical applications of ILs done by Moniruzzaman et al. [15] showed the benefits of this particular combination. Added to this is the possibility for completely replacing surfactants with ILs, which can act as the surfactant within the system. Vesicle and IL combination is also quite versatile as discussed for surfactants. ILs can be assembled into vesicles with great ease and even act as actives that assemble into vesicles [95,96]. When employing this approach, however, care must be taken when choosing the IL, as the choice of IL will affect bioactivity greatly.

The application of ILs in these systems will greatly enhance their activity, safety, and overall therapeutic outcomes. The above listed incorporations are all feasible with current technologies and are limited only by the diversity of ILs, making rational selection of appropriate combinations extremely difficult. While applications of ILs in other industrial sectors are growing rapidly, the stagnant medical applications are not simply due to research; scarce safety data also decrease the likelihood that formulations with ILs will make it to the open market. This issue can be overcome by conducting more in-depth cytotoxicity cell studies, and in vivo and ex vivo studies, which could lead to the translation of ILs from research applications to the development of drug delivery systems.

Table 3. Presented here are some of the applications of ILs synergistically used to enhance drug delivery with chemical permeation enhancers (CPEs). APIs—active pharmaceutical ingredients.

Chemical Enhancer	ILs	Synergism Documented	Reference
Lipid vesicles	ILs based on methylimidazolium chloride	1) Enhanced permeation 2) Improved therapeutic range 3) Vesicle stabilized	[97]
Surfactants	Dimethyl-imidazolium dimethyl-phosphate	1) Enhanced permeation of sparingly soluble APIs 2) Reduced cytotoxicity	[98]
Terpenes	Menthoxymethyl-3-methylimidazolium chloride	1) Enhanced spatial drug delivery 2) Improved targeting of specified receptors	[99]
Amines	Amine-based ILs	1) Permeation of hydrophobic and hydrophilic molecules 2) No skin tissue injury during permeation	[81]
Alcohols	*N-tert*-Alcohol-substituted imidazole	1) Medium for selective reactions 2) Lowered instances of unwanted byproducts	[100]

4.3. ILs in Drug Dissolution for Improving Bioavailability in TDDS

4.3.1. Drug Moiety and Delivery System Modification

As mentioned earlier, the use of ILs as reaction media and solvents is well studied. They can be modified to achieve any number of favorable properties; however, this is not the limit as they can be further used in API synthesis to aid as catalysts and transformative media [101]. Particularly when using biotechnological processes, ILs can act as selective enzyme enablers to improve yields [4] and limit waste produced. The nano ultra-structures (as discussed earlier) provide an ideal environment for selective reactions to favorably alter APIs. While their use as catalysts and solvents helps reduce waste produced, they found a niche when applied to enzymes. They can help with stability, enhance the activity of enzymes, control folding in proteins, and reduce aggregation of the proteins [14]. Most ILs contain some degree of hydration and can further adsorb water and can affect the stability of these macromolecules.

ILs were applied to modify current emerging delivery systems for effective delivery of APIs. These include ILs acting as functionalizing agents, solvents, dispersing agents, nanoparticles, nanocarriers, and substrate activation. Emulsion modification became a key strategy, as evidenced in work by Kandasamy et al. [89], where ammonium acetate-based ILs promote the ease of formulation of IL-in-oil microemulsions, further improving the stability of the system and enhancing solubilization. This was also evidenced in work done by Yoshiura et al. [88], demonstrating microemulsion size and size distributions are improved by IL incorporation, leading to enhanced transdermal permeation. ILs in these microemulsion-based systems act as solvents, whereby they can solubilize the API. Further to this, they can replace both the hydrophobic and hydrophilic moieties in microemulsion-based systems. The advantages of IL applications in these systems include the solvating power of ILs, which require less solvent and can solubilize a wide variety of hydrophilic and hydrophobic drugs; their use as permeation enhancers is a key feature in transdermal systems [80].

4.3.2. Forming a Complex of IL with the API

ILs can be implemented in a variety of ways to aid in formulation studies. A common strategy to improve absorption and bioavailability is that APIs can be modified and formulated as prodrugs. Prodrugs are inactive biologically until they are bio-transformed into the active metabolite. Advantages of prodrug platforms include increased bioavailability, lowered metabolism, improved site specificity, and controlled release of the active metabolite. ILs were combined with APIs to make prodrug platforms [83] via the addition of hydrolysable ionic groups to neutrally charged APIs. These were then combined with the correct counterions to formulate a new prodrug IL. The advantages to this strategy are (a) the use of a counterion that is also an API leading to a dual-function IL, and (b) combination with a penetration enhancer in transdermal systems to improve permeation [102,103].

The prodrug formulation can further be enhanced by combining IL-APIs with these prodrug platforms, thereby conferring the advantages of both strategies. Cojocaru et al. successfully created prodrug API-ILs by adding hydrolysable functional groups to neutral APIs that can be paired with suitable anion counterparts [104]. The authors of the paper did acknowledge that the chosen IL forms used may not be the most suitable and encouraged substitution with those that are more suitable such as carboxylic acids and amines. By using this as a starting point, the advantages can be numerous. Advantages of prodrugs and ILs aside, the pairing of synergistic drugs with appropriate anion and cation pairs (dual APIs discussed later) can lead to the need for less individual medicaments, thereby further enhancing patient compliance.

Release studies conducted by Cojocaru et al. [104] found that newly developed acetaminophen-based prodrugs had slower release times (at 210 min ~89% release) when compared to the neutral forms (at 210 min >97% release). This slower release profile is not in itself a disadvantage, particularly when considering extended-release systems are becoming more common.

4.3.3. Dissolving the API in the IL (Solvation)

Solvation of an API forms the first stages of drug absorption and is a critical factor to consider in any efficient drug delivery system. Currently, the use of organic solvents in the pharmaceutical industry is an undeniable challenge to sustainable growth in this sector. Further to this is the organic contamination of pharmaceutical products from these solvents. Alternative solvents and strategies are long overdue, in an effort to produce greener and more sustainable practices with lowered waste generation (corresponding e-factors). As such, ILs provide a suitable alternative to volatile organic solvents for a variety of applications, including use as a reaction medium. The advantages they offer are numerous and include improved solubilization [105], faster reaction rates, higher purity end products, decreased waste generation, and less fastidious reaction conditions. The preferred cation in these reaction media is methylimidazole combined with anions such as BF_4, PF_6, and NTf_2. This combination and other ILs can be used in the preparation of a variety of drugs, including antiviral agents, non-steroidal anti-inflammatory drugs (NSAIDs), anti-neoplastic agents, and anti-infectives.

For an API-solvent combination to be used in medicines and bio-medical applications, its solubility must be greater than 1 mg/mL. Many IL candidates may exist meeting this standard for any given API, but their toxicity profiles remain to be investigated [14]. This is a significant impediment to the widespread use of ILs in medical and biomedical fields. Limited toxicity studies were conducted, and much more work needs to be done in this field, as can be seen by in-depth meta-analysis studies such as those conducted by Heckenbach and co-workers [106], where many studies dealing with ILs (~39,000) were reported on, but a significantly lower number (~220) deal with toxicity in ILs. By using biologically derived bases, the compatibility can be improved, and the toxicity can be limited [94]. Bioinspired ILs have limited issues with toxicity as they can be metabolized and excreted; further modifications such as alkyl group substitution with ester groups further reduce toxicity. IL blends were also used to accommodate and alter varying solubility standards. Using water-miscible ILs in combination with hydrophobic ILs allows a degree of freedom when considering hydrophobic APIs.

4.4. ILs as Active Pharmaceutical Ingredients (APIs) in TDDS

APIs in large part exist as crystalline solids due to historical reasons, better stability, and support from guidelines such as the Food and Drug Administration (FDA) guidelines, which give all the necessary information for solid crystalline API analytical procedures, and decision trees providing streamlined manufacturing processes. This is not the same for other API forms. The disadvantages associated with these crystalline solid APIs affect many physicochemical properties and, ultimately, the behavior of these compounds when introduced in medicaments. The most apparent and problematic challenge is polymorphism [14]. Polymorphs occur when crystalline substances crystalize in more than a single form. This can lead to challenges in solubility and absorption, ultimately affecting bioavailability. Polymorph forms can be affected by manufacturing conditions such as solvent choice, temperature, pressure, and mechanical stress. Polymorphs do not only occur during manufacture, but also during long term storage, where less stable polymorphs will revert to more stable polymorphic structures under storage conditions. While stable polymorphs are less prone to degradation, they are often more difficult to solubilize and, when dissolved, they can salt out.

Challenges such as polymorphism, solubility, and bioavailability can be overcome by using ILs [14]. Historically, the first API-ILs were considered to be miconazole derivatives [4]. Building on this work, the formation of ranitidine docusate, lidocaine docusate, and dual API-ILs was later reported. These studies led to (a) the synthesis of didecyldimethylammonium bromide and sodium ibuprofen to give didecyldimethylammoniumibuprofenate, and (b) a demonstration that the API could be either the anion [107], cation [108], or both [82]. Work done to date (pertinent examples illustrated in Table 4 include the use of lidocaine, sulfacetamide, ibuprofen, cinnamic acid, piperacillin, acetyl-salicylic acid, and penicillin G [14]. The use of IL-APIs not only helps overcome manufacturing challenges, but also positively impacts therapeutic outcomes by improving API bioavailability and penetration, as well as providing alternative delivery methods and beneficial synergistic interactions.

Table 4. The use of IL-APIs with beneficial delivery and therapeutic outcomes.

IL-API Formed	Synergism	Efficacy	Reference
Acetyl salicylic acid/salicylate	Improved manufacturing methods Altered side-effect Profile	Solvent-free synthesis Lowered Gastric distress	[14,109–111]
Lidocaine docusate	Improved therapeutic outcome	Longer duration of action	[107,112]
Ranitidine docusate	Improved manufacture outcomes	Improved polymorphic challenges	[107]
Didecyldimethylammonium ibuprofenate	Proof of concept	Dual API formation	[107]

5. Proposed/Future Opportunities for Synergism in TDDS

The incorporation of ILs in the abovementioned systems can be summarized to a few key strategies for future deployment, as follows:

1) The use of ILs in a pre-treatment cycle: By pre-treating with ILs, the skin barrier properties are decreased. This in turn would allow more of the active compounds being employed with physical permeation to penetrate.
2) Using ILs in the permeation system: ILs are a diverse medium in which permeation systems can be made and modified. They allow incorporation into systems by acting as electrolyte solutions, coating media, and media in which delivery systems can be synthesized. Care must be taken to ensure that unwanted products do not form when changing the manufacturing process.
3) The use of ILs as a bioactive: Certain drug molecules can be derivatized as ILs and can be incorporated into other permeation systems. This will improve permeation and remove unwanted polymorphs. Using prodrug platforms in this method may enhance permeation; however, the need for the molecule to be activated may lead to longer therapeutic lead times.
4) Synergistic combination: A primary method of improving permeation with chemical enhancers is the combination use of ILs with another chemical enhancer, which can lead to synergistic or additive penetration. They can be combined in a single formulation where the IL can be the primary solvent, co-solvent, or surfactant, or can be employed as a second permeation enhancer. The risk to this strategy is that the irritation may be much more apparent during patient use.
5) Altered favorable environment for drug molecules: The use of ILs as a medium for solubilizing drug molecules can be varied to be more favorable. This strategy may not be applicable to a large number of drug molecules, as most are synthesized as crystalline solids as opposed to liquids. These classes of drug molecules can therefore be easily incorporated into the ILs.
6) Diverse activity profile: ILs can act as a number of formulation components. They can act as vehicles solubilizing drug molecules, surfactants, micelles, and permeation enhancers, and they can replace aqueous or oily phases in formulations. This diverse portfolio makes them ideal for incorporation into many formulations. However, a lack of long-term toxicity studies limits the widespread use of ILs.
7) Modification of embedded substrates: The use of ILs to enhance drug profiles has sound evidence. These are often favorable for drug delivery where drug molecular profiles are altered to enhance permeation and systemic absorption. The long-term stability and safety of these altered drug molecules is largely unknown and, therefore, requires much work before this strategy can reach clinical applications.

6. Conclusions

The benefits of transdermal drug delivery are apparent, and a significant number of potential techniques already exist. However, systemic absorption of most drug molecules is still elusive due to the skin and its barrier properties. The possibility of using ILs as transdermal delivery systems or permeation enhancers, and even in synergistic combinations without the risk of major toxicity is a trademark of their versatile nature. They can be combined in a variety of ways with existing permeation systems. ILs will clearly have a large role to play in transdermal drug delivery in the near future. Their use as singular agents remains to be proven; however, when combined with other formulation strategies, the activity enhancing transdermal permeation is remarkably greater.

Funding: This work was funded by the National Research Foundation (NRF) of South Africa.

Conflicts of Interest: The authors declare no conflicts of interest.

References

1. Gadilohar, B.L.; Shankarling, G.S. Choline based ionic liquids and their applications in organic transformation. *J. Mol. Liquids* **2017**, *227*, 234–261. [CrossRef]
2. Bermudez, M.D.; Jimenez, A.E.; Sanes, J.; Carrion, F.J. Ionic liquids as advanced lubricant fluids. *Molecules* **2009**, *14*, 2888–2908. [CrossRef] [PubMed]
3. Das, R.N.; Roy, K. Advances in qspr/qstr models of ionic liquids for the design of greener solvents of the future. *Mol. Divers.* **2013**, *17*, 151–196. [CrossRef] [PubMed]
4. Egorova, K.S.; Gordeev, E.G.; Ananikov, V.P. Biological activity of ionic liquids and their application in pharmaceutics and medicine. *Chem. Rev.* **2017**, *117*, 7132–7189. [CrossRef]
5. Amiril, S.A.S.; Rahim, E.A.; Syahrullail, S. A review on ionic liquids as sustainable lubricants in manufacturing and engineering: Recent research, performance, and applications. *J. Clean. Prod.* **2017**, *168*, 1571–1589. [CrossRef]
6. Dong, K.; Liu, X.; Dong, H.; Zhang, X.; Zhang, S. Multiscale studies on ionic liquids. *Chem. Rev.* **2017**, *117*, 6636–6695. [CrossRef]
7. Lim, G.S.; Jaenicke, S.; Klahn, M. How the spontaneous insertion of amphiphilic imidazolium-based cations changes biological membranes: A molecular simulation study. *Phys. Chem. Chem. Phys.* **2015**, *17*, 29171–29183. [CrossRef] [PubMed]
8. Kundu, N.; Roy, S.; Mukherjee, D.; Maiti, T.K.; Sarkar, N. Unveiling the interaction between fatty-acid-modified membrane and hydrophilic imidazolium-based ionic liquid: Understanding the mechanism of ionic liquid cytotoxicity. *J. Phys. Chem. B* **2017**, *121*, 8162–8170. [CrossRef]
9. Ferraz, R.; Branco, L.C.; Prudencio, C.; Noronha, J.P.; Petrovski, Z. Ionic liquids as active pharmaceutical ingredients. *ChemMedChem* **2011**, *6*, 975–985. [CrossRef]
10. Shamshina, J.L.; Barber, P.S.; Rogers, R.D. Ionic liquids in drug delivery. *Expert Opin. Drug Deliv.* **2013**, *10*, 1367–1381. [CrossRef]
11. Dobler, D.; Schmidts, T.; Klingenhofer, I.; Runkel, F. Ionic liquids as ingredients in topical drug delivery systems. *Int. J. Pharm.* **2013**, *441*, 620–627. [CrossRef] [PubMed]
12. Marrucho, I.M.; Branco, L.C.; Rebelo, L.P. Ionic liquids in pharmaceutical applications. *Annu. Rev. Chem. Biomol. Eng.* **2014**, *5*, 527–546. [CrossRef] [PubMed]
13. Shamshina, J.L.; Kelley, S.P.; Gurau, G.; Rogers, R.D. Chemistry: Develop ionic liquid drugs. *Nature* **2015**, *528*, 188–189. [CrossRef]
14. Adawiyah, N.; Moniruzzaman, M.; Hawatulailaa, S.; Goto, M. Ionic liquids as a potential tool for drug delivery systems. *MedChemComm* **2016**, *7*, 1881–1897. [CrossRef]
15. Moniruzzaman, M.; Tahara, Y.; Tamura, M.; Kamiya, N.; Goto, M. Ionic liquid-assisted transdermal delivery of sparingly soluble drugs. *Chem. Commun.* **2010**, *46*, 1452–1454. [CrossRef] [PubMed]
16. Sivapragasam, M.; Moniruzzaman, M.; Goto, M. Recent advances in exploiting ionic liquids for biomolecules: Solubility, stability and applications. *Biotechnol. J.* **2016**, *11*, 1000–1013. [CrossRef] [PubMed]
17. Goindi, S.; Kaur, R.; Kaur, R. An ionic liquid-in-water microemulsion as a potential carrier for topical delivery of poorly water soluble drug: Development, ex-vivo and in-vivo evaluation. *Int. J. Pharm.* **2015**, *495*, 913–923. [CrossRef]
18. Mbous, Y.P.; Hayyan, M.; Hayyan, A.; Wong, W.F.; Hashim, M.A.; Looi, C.Y. Applications of deep eutectic solvents in biotechnology and bioengineering-promises and challenges. *Biotechnol. Adv.* **2017**, *35*, 105–134. [CrossRef]
19. Troter, D.Z.; Todorovic, Z.B.; Dokic-Stojanovic, D.R.; Stamenkovic, O.S.; Veljkovic, V.B. Application of ionic liquids and deep eutectic solvents in biodiesel production: A review. *Renew. Sustain. Energy Rev.* **2016**, *61*, 473–500. [CrossRef]
20. Kudlak, B.; Owczarek, K.; Namiesnik, J. Selected issues related to the toxicity of ionic liquids and deep eutectic solvents—A review. *Environ. Sci. Pollut. Res. Int.* **2015**, *22*, 11975–11992. [CrossRef]
21. Wang, B.; Qin, L.; Mu, T.; Xue, Z.; Gao, G. Are ionic liquids chemically stable? *Chem. Rev.* **2017**, *117*, 7113–7131. [CrossRef] [PubMed]
22. Lei, Z.; Chen, B.; Koo, Y.M.; MacFarlane, D.R. Introduction: Ionic liquids. *Chem. Rev.* **2017**, *117*, 6633–6635. [CrossRef] [PubMed]

23. Marsh, K.N.; Boxall, J.A.; Lichtenthaler, R. Room temperature ionic liquids and their mixtures—A review. *Fluid Phase Equilibria* **2004**, *219*, 93–98. [CrossRef]
24. Earle, M.J.; Seddon, K.R. Ionic liquids. Green solvents for the future. *Pure Appl. Chem.* **2000**, *72*, 1391–1398. [CrossRef]
25. Agatemor, C.; Ibsen, K.N.; Tanner, E.E.L.; Mitragotri, S. Ionic liquids for addressing unmet needs in healthcare. *Bioeng. Transl. Med.* **2018**, *3*, 7–25. [CrossRef]
26. Boudalis, A.K.; Rogez, G.; Heinrich, B.; Raptis, R.G.; Turek, P. Towards ionic liquids with tailored magnetic properties: Bmim(+) salts of ferro- and antiferromagnetic cu triangles. *Dalton Trans.* **2017**, *46*, 12263–12273. [CrossRef]
27. Hayes, R.; Warr, G.G.; Atkin, R. Structure and nanostructure in ionic liquids. *Chem. Rev.* **2015**, *115*, 6357–6426. [CrossRef]
28. Visser, A.E.; Reichert, W.M.; Swatloski, R.P.; Willauer, H.D.; Huddleston, J.G.; Rogers, R.D. Characterization of hydrophilic and hydrophobic ionic liquids: Alternatives to volatile organic compounds for liquid-liquid separations. In *Ionic Liquids*; American Chemical Society: Washington, DC, USA, 2002; Volume 818, pp. 289–308.
29. Cao, Y.; Yao, S.; Wang, X.; Peng, Q.; Song, H. The physical and chemical properties of ionic liquids and its application in extraction. In *Handbook of Ionic Liquids: Properties, Applications and Hazards*; Mun, J., Sim, H., Eds.; Nova Science Publishers: New York, NY, USA, 2012; pp. 145–172.
30. Tokuda, H.; Hayamizu, K.; Ishii, K.; Susan, M.A.; Watanabe, M. Physicochemical properties and structures of room temperature ionic liquids. 2. Variation of alkyl chain length in imidazolium cation. *J. Phys. Chem. B* **2005**, *109*, 6103–6110. [CrossRef]
31. Bhargava, B.L.; Klein, M.L. Initial stages of aggregation in aqueous solutions of ionic liquids: Molecular dynamics studies. *J. Phys. Chem. B* **2009**, *113*, 9499–9505. [CrossRef]
32. Miskiewicz, A.; Ceranowicz, P.; Szymczak, M.; Bartus, K.; Kowalczyk, P. The use of liquids ionic fluids as pharmaceutically active substances helpful in combating nosocomial infections induced by klebsiella pneumoniae new delhi strain, acinetobacter baumannii and enterococcus species. *Int. J. Mol. Sci.* **2018**, *19*, 2779. [CrossRef]
33. Gao, W.W.; Zhang, F.X.; Zhang, G.X.; Zhou, C.H. Key factors affecting the activity and stability of enzymes in ionic liquids and novel applications in biocatalysis. *Biochem. Eng. J.* **2015**, *99*, 67–84. [CrossRef]
34. Mantz, R.A.; Trulove, P. *Viscosity and Density of Ionic Liquids*; Wiley-VCH Verlag GmbH & Co: Weinheim, Germany, 2003; Volume 2, pp. 56–68.
35. Katritzky, A.R.; Lomaka, A.; Petrukhin, R.; Jain, R.; Karelson, M.; Visser, A.E.; Rogers, R.D. Qspr correlation of the melting point for pyridinium bromides, potential ionic liquids. *J. Chem. Inf. Comput. Sci.* **2002**, *42*, 71–74. [CrossRef] [PubMed]
36. Takekiyo, T.; Yoshimura, Y. Suppression and dissolution of amyloid aggregates using ionic liquids. *Biophys. Rev.* **2018**, *10*, 853–860. [CrossRef] [PubMed]
37. Pillai, V.V.S.; Benedetto, A. Ionic liquids in protein amyloidogenesis: A brief screenshot of the state-of-the-art. *Biophys. Rev.* **2018**, *10*, 847–852. [CrossRef] [PubMed]
38. Benedetto, A. Room-temperature ionic liquids meet bio-membranes: The state-of-the-art. *Biophys. Rev.* **2017**, *9*, 309–320. [CrossRef] [PubMed]
39. Ibsen, K.; Ma, H.; Banerjee, A.; Tanner, E.; Nangia, S.; Mitragotri, S. *Mechanism of Antibacterial Activity of Choline-Based Ionic Liquids*; American Chemical Society: Washington, DC, USA, 2018; Volume 4.
40. Zakrewsky, M.; Lovejoy, K.S.; Kern, T.L.; Miller, T.E.; Le, V.; Nagy, A.; Goumas, A.M.; Iyer, R.S.; Del Sesto, R.E.; Koppisch, A.T.; et al. Ionic liquids as a class of materials for transdermal delivery and pathogen neutralization. *Proc. Natl. Acad. Sci. USA* **2014**, *111*, 13313–13318. [CrossRef] [PubMed]
41. Wang, D.; Richter, C.; Ruhling, A.; Drucker, P.; Siegmund, D.; Metzler-Nolte, N.; Glorius, F.; Galla, H.J. A remarkably simple class of imidazolium-based lipids and their biological properties. *Chemistry* **2015**, *21*, 15123–15126. [CrossRef] [PubMed]
42. Wang, D.; Richter, C.; Ruhling, A.; Huwel, S.; Glorius, F.; Galla, H.J. Anti-tumor activity and cytotoxicity in vitro of novel 4,5-dialkylimidazolium surfactants. *Biochem. Biophys. Res. Commun.* **2015**, *467*, 1033–1038. [CrossRef] [PubMed]
43. Benedetto, A.; Bodo, E.; Gontrani, L.; Ballone, P.; Caminiti, R. Amino acid anions in organic ionic compounds. An ab initio study of selected ion pairs. *J. Phys. Chem. B* **2014**, *118*, 2471–2486. [CrossRef]

44. Evans, K.O. Supported phospholipid membrane interactions with 1-butyl-3-methylimidazolium chloride. *J. Phys. Chem. B* **2008**, *112*, 8558–8562. [CrossRef]
45. Jing, B.; Lan, N.; Qiu, J.; Zhu, Y. Interaction of ionic liquids with a lipid bilayer: A biophysical study of ionic liquid cytotoxicity. *J. Phys. Chem. B* **2016**, *120*, 2781–2789. [CrossRef] [PubMed]
46. Benedetto, A.; Heinrich, F.; Gonzalez, M.A.; Fragneto, G.; Watkins, E.; Ballone, P. Structure and stability of phospholipid bilayers hydrated by a room-temperature ionic liquid/water solution: A neutron reflectometry study. *J. Phys. Chem. B* **2014**, *118*, 12192–12206. [CrossRef] [PubMed]
47. Benedetto, A.; Bingham, R.J.; Ballone, P. Structure and dynamics of popc bilayers in water solutions of room temperature ionic liquids. *J. Chem. Phys.* **2015**, *142*, 124706. [CrossRef] [PubMed]
48. Drucker, P.; Ruhling, A.; Grill, D.; Wang, D.; Draeger, A.; Gerke, V.; Glorius, F.; Galla, H.J. Imidazolium salts mimicking the structure of natural lipids exploit remarkable properties forming lamellar phases and giant vesicles. *Langmuir* **2017**, *33*, 1333–1342. [CrossRef] [PubMed]
49. Prausnitz, M.R.; Langer, R. Transdermal drug delivery. *Nat. Biotechnol.* **2008**, *26*, 1261–1268. [CrossRef] [PubMed]
50. Wang, Q.L.; Zhu, D.D.; Liu, X.B.; Chen, B.Z.; Guo, X.D. Microneedles with controlled bubble sizes and drug distributions for efficient transdermal drug delivery. *Sci. Rep.* **2016**, *6*, 38755. [CrossRef] [PubMed]
51. Kabashima, K.; Honda, T.; Ginhoux, F.; Egawa, G. The immunological anatomy of the skin. *Nat. Rev. Immunol* **2019**, *19*, 19–30. [CrossRef] [PubMed]
52. Lee, S.H.; Jeong, S.K.; Ahn, S.K. An update of the defensive barrier function of skin. *Yonsei Med. J.* **2006**, *47*, 293–306. [CrossRef] [PubMed]
53. Matsui, T.; Amagai, M. Dissecting the formation, structure and barrier function of the stratum corneum. *Int. Immunol.* **2015**, *27*, 269–280. [CrossRef]
54. Karande, P.; Jain, A.; Mitragotri, S. Discovery of transdermal penetration enhancers by high-throughput screening. *Nat. Biotechnol.* **2004**, *22*, 192–197. [CrossRef]
55. Rastogi, V.; Yadav, P. Transdermal drug delivery system: An overview. *Asian J. Pharm.* **2012**, *6*, 161–170. [CrossRef]
56. Ita, K. Ceramic microneedles and hollow microneedles for transdermal drug delivery: Two decades of research. *J. Drug Deliv. Sci. Technol.* **2018**, *44*, 314–322. [CrossRef]
57. Naik, A.; Kalia, Y.N.; Guy, R.H. Transdermal drug delivery: Overcoming the skin's barrier function. *Pharm. Sci. Technol. Today* **2000**, *3*, 318–326. [CrossRef]
58. Wickett, R.R.; Visscher, M.O. Structure and function of the epidermal barrier. *Am. J. Infect. Control* **2006**, *34*, S98–S110. [CrossRef]
59. Masukawa, Y.; Narita, H.; Shimizu, E.; Kondo, N.; Sugai, Y.; Oba, T.; Homma, R.; Ishikawa, J.; Takagi, Y.; Kitahara, T.; et al. Characterization of overall ceramide species in human stratum corneum. *J. Lipid Res.* **2008**, *49*, 1466–1476. [CrossRef] [PubMed]
60. Maibach, H. Dermatological formulations: Percutaneous absorption. By brian w. Barry. Marcel dekker, 270 madison avenue, new york, ny 10016. 1983. 479pp. 16 × 23.5cm. Price $55.00 (2070 higher outside the us. And canada). *J. Pharm. Sci.* **1984**, *73*, 573. [CrossRef]
61. Williams, A.C.; Barry, B.W. Penetration enhancers. *Adv. Drug Deliv. Rev.* **2012**, *64*, 128–137. [CrossRef]
62. Karande, P.; Jain, A.; Ergun, K.; Kispersky, V.; Mitragotri, S. Design principles of chemical penetration enhancers for transdermal drug delivery. *Proc. Natl. Acad. Sci. USA* **2005**, *102*, 4688–4693. [CrossRef] [PubMed]
63. Munch, S.; Wohlrab, J.; Neubert, R.H.H. Dermal and transdermal delivery of pharmaceutically relevant macromolecules. *Eur. J. Pharm. Biopharm.* **2017**, *119*, 235–242. [CrossRef] [PubMed]
64. Potts, R.O.; Guy, R.H. Predicting skin permeability. *Pharm. Res.* **1992**, *9*, 663–669. [CrossRef] [PubMed]
65. Wang, Y.; Vlasova, A.; Velasquez, D.E.; Saif, L.J.; Kandasamy, S.; Kochba, E.; Levin, Y.; Jiang, B. Skin vaccination against rotavirus using microneedles: Proof of concept in gnotobiotic piglets. *PLoS ONE* **2016**, *11*, e0166038. [CrossRef] [PubMed]
66. Chen, Y.; Quan, P.; Liu, X.; Wang, M.; Fang, L. Novel chemical permeation enhancers for transdermal drug delivery. *Asian J. Pharm. Sci.* **2014**, *9*, 51–64. [CrossRef]
67. Arya, J.; Prausnitz, M.R. Microneedle patches for vaccination in developing countries. *J. Control. Release* **2016**, *240*, 135–141. [CrossRef] [PubMed]

68. Liu, S.; Jin, M.N.; Quan, Y.S.; Kamiyama, F.; Katsumi, H.; Sakane, T.; Yamamoto, A. The development and characteristics of novel microneedle arrays fabricated from hyaluronic acid, and their application in the transdermal delivery of insulin. *J. Control. Release* **2012**, *161*, 933–941. [CrossRef] [PubMed]
69. Latsch, S.; Selzer, T.; Fink, L.; Kreuter, J. Crystallisation of estradiol containing tdds determined by isothermal microcalorimetry, x-ray diffraction, and optical microscopy. *Eur. J. Pharm. Biopharm.* **2003**, *56*, 43–52. [CrossRef]
70. Kogan, A.; Garti, N. Microemulsions as transdermal drug delivery vehicles. *Adv. Colloid Interface Sci.* **2006**, *123–126*, 369–385. [CrossRef] [PubMed]
71. Milewski, M.; Yerramreddy, T.R.; Ghosh, P.; Crooks, P.A.; Stinchcomb, A.L. In vitro permeation of a pegylated naltrexone prodrug across microneedle-treated skin. *J. Control. Release* **2010**, *146*, 37–44. [CrossRef] [PubMed]
72. Miyagi, T.; Hikima, T.; Tojo, K. Effect of molecular weight of penetrants on iontophoretic transdermal delivery in vitro. *J. Chem. Eng. Jpn.* **2006**, *39*, 360–365. [CrossRef]
73. Banga, A.K.; Bose, S.; Ghosh, T.K. Iontophoresis and electroporation: Comparisons and contrasts. *Int. J. Pharm.* **1999**, *179*, 1–19. [CrossRef]
74. Kumar, S.K.; Bhowmik, D.; Komala, M. Transdermal sonophoresis technique-an approach for controlled drug delivery. *Indian J. Res. Pharm. Biotechnol.* **2013**, *1*, 379–381.
75. Prausnitz, M.R. Microneedles for transdermal drug delivery. *Adv. Drug Deliv. Rev.* **2004**, *56*, 581–587. [CrossRef] [PubMed]
76. Lane, M.E. Skin penetration enhancers. *Int. J. Pharm.* **2013**, *447*, 12–21. [CrossRef]
77. Tfayli, A.; Guillard, E.; Manfait, M.; Baillet-Guffroy, A. Molecular interactions of penetration enhancers within ceramides organization: A raman spectroscopy approach. *Analyst* **2012**, *137*, 5002–5010. [CrossRef] [PubMed]
78. Barry, B.W. *Dermatological Formulations: Percutaneous Absorption*; Marcel Dekker: New York, NY, USA, 1983.
79. Karande, P.; Mitragotri, S. Enhancement of transdermal drug delivery via synergistic action of chemicals. *Biochim. Biophys. Acta* **2009**, *1788*, 2362–2373. [CrossRef] [PubMed]
80. Wang, C.; Zhu, J.; Zhang, D.; Yang, Y.; Zheng, L.; Qu, Y.; Yang, X.; Cui, X. Ionic liquid—Microemulsions assisting in the transdermal delivery of dencichine: Preparation, in-vitro and in-vivo evaluations, and investigation of the permeation mechanism. *Int. J. Pharm.* **2018**, *535*, 120–131. [CrossRef] [PubMed]
81. Kubota, K.; Shibata, A.; Yamaguchi, T. The molecular assembly of the ionic liquid/aliphatic carboxylic acid/aliphatic amine as effective and safety transdermal permeation enhancers. *Eur. J. Pharm. Sci.* **2016**, *86*, 75–83. [CrossRef] [PubMed]
82. Park, H.J.; Prausnitz, M.R. Lidocaine-ibuprofen ionic liquid for dermal anesthesia. *AICHE J.* **2015**, *61*, 2732–2738. [CrossRef]
83. Zakrewsky, M.; Mitragotri, S. Therapeutic rnai robed with ionic liquid moieties as a simple, scalable prodrug platform for treating skin disease. *J. Control. Release* **2016**, *242*, 80–88. [CrossRef]
84. Monti, D.; Egiziano, E.; Burgalassi, S.; Chetoni, P.; Chiappe, C.; Sanzone, A.; Tampucci, S. Ionic liquids as potential enhancers for transdermal drug delivery. *Int. J. Pharm.* **2017**, *516*, 45–51. [CrossRef]
85. Zhang, D.; Wang, H.J.; Cui, X.M.; Wang, C.X. Evaluations of imidazolium ionic liquids as novel skin permeation enhancers for drug transdermal delivery. *Pharm. Dev. Technol.* **2017**, *22*, 511–520. [CrossRef]
86. Banerjee, A.; Ibsen, K.; Iwao, Y.; Zakrewsky, M.; Mitragotri, S. *Transdermal Protein Delivery Using Choline and Geranate (cage) Deep Eutectic Solvent*; WILEY-VCH Verlag GmbH & Co. KGaA: Weinheim, Germany, 2017; Volume 6.
87. Araki, S.; Wakabayashi, R.; Moniruzzaman, M.; Kamiya, N.; Goto, M. Ionic liquid-mediated transcutaneous protein delivery with solid-in-oil nanodispersions. *MedChemComm* **2015**, *6*, 2124–2128. [CrossRef]
88. Yoshiura, H.; Tamura, M.; Aso, M.; Kamiya, N.; Goto, M. *Ionic Liquid-in-Oil Microemulsions as Potential Carriers for the Transdermal Delivery of Methotrexate*; The Society of Chemical Engineers: Tokyo, Japan, 2013; Volume 46, pp. 794–796.
89. Kandasamy, S.; Moniruzzaman, M.; Sivapragasam, M.; Shamsuddin, M.R.; Mutalib, M.I.A. Formulation and characterization of acetate based ionic liquid in oil microemulsion as a carrier for acyclovir and methotrexate. *Sep. Purif. Technol.* **2018**, *196*, 149–156. [CrossRef]
90. Aboofazeli, R.; Zia, H.; Needham, T.E. Transdermal delivery of nicardipine: An approach to in vitro permeation enhancement. *Drug Deliv.* **2002**, *9*, 239–247. [CrossRef] [PubMed]

91. Yoo, B.; Jing, B.; Jones, S.E.; Lamberti, G.A.; Zhu, Y.; Shah, J.K.; Maginn, E.J. Molecular mechanisms of ionic liquid cytotoxicity probed by an integrated experimental and computational approach. *Sci. Rep.* **2016**, *6*, 19889. [CrossRef] [PubMed]
92. Frade, R.F.; Afonso, C.A. Impact of ionic liquids in environment and humans: An overview. *Hum. Exp. Toxicol.* **2010**, *29*, 1038–1054. [CrossRef] [PubMed]
93. Hou, X.D.; Liu, Q.P.; Smith, T.J.; Li, N.; Zong, M.H. Evaluation of toxicity and biodegradability of cholinium amino acids ionic liquids. *PLoS ONE* **2013**, *8*, e59145. [CrossRef]
94. Caparica, R.; Julio, A.; Baby, A.R.; Araujo, M.E.M.; Fernandes, A.S.; Costa, J.G.; Santos de Almeida, T. Choline-amino acid ionic liquids as green functional excipients to enhance drug solubility. *Pharmaceutics* **2018**, *10*, 288. [CrossRef]
95. Zhang, L.; Liu, J.; Tian, T.; Gao, Y.; Ji, X.; Li, Z.; Luan, Y. Pharmaceutically active ionic liquid self-assembled vesicles for the application as an efficient drug delivery system. *ChemPhysChem* **2013**, *14*, 3454–3457. [CrossRef]
96. Rao, K.S.; So, S.; Kumar, A. Vesicles and reverse vesicles of an ionic liquid in ionic liquids. *Chem. Commun.* **2013**, *49*, 8111–8113. [CrossRef]
97. Hanna, S.L.; Huang, J.L.; Swinton, A.J.; Caputo, G.A.; Vaden, T.D. Synergistic effects of polymyxin and ionic liquids on lipid vesicle membrane stability and aggregation. *Biophys. Chem.* **2017**, *227*, 1–7. [CrossRef]
98. Moniruzzaman, M.; Tamura, M.; Tahara, Y.; Kamiya, N.; Goto, M. Ionic liquid-in-oil microemulsion as a potential carrier of sparingly soluble drug: Characterization and cytotoxicity evaluation. *Int. J. Pharm.* **2010**, *400*, 243–250. [CrossRef] [PubMed]
99. Flieger, J.; Feder-Kubis, J.; Tatarczak-Michalewska, M.; Plazinska, A.; Madejska, A.; Swatko-Ossor, M. Natural terpene derivatives as new structural task-specific ionic liquids to enhance the enantiorecognition of acidic enantiomers on teicoplanin-based stationary phase by high-performance liquid chromatography. *J. Sep. Sci.* **2017**, *40*, 2374–2381. [CrossRef] [PubMed]
100. Shinde, S.S.; Lee, B.S.; Chi, D.Y. Synergistic effect of two solvents, tert-alcohol and ionic liquid, in one molecule in nucleophilic fluorination. *Org. Lett.* **2008**, *10*, 733–735. [CrossRef] [PubMed]
101. Mansoor, S.S.; Aswin, K.; Logaiya, K.; Sudhan, S.P.N. [bmim]bf4 ionic liquid: An efficient reaction medium for the one-pot multi-component synthesis of 2-amino-4, 6-diphenylpyridine-3-carbonitrile derivatives. *J. Saudi Chem. Soc.* **2016**, *20*, 517–522. [CrossRef]
102. Davis, J.H., Jr.; Fox, P.A. From curiosities to commodities: Ionic liquids begin the transition. *Chem. Commun.* **2003**, *11*, 1209–1212. [CrossRef]
103. Nishi, N.; Kawakami, T.; Shigematsu, F.; Yamamoto, M.; Kakiuchi, T. Fluorine-free and hydrophobic room-temperature ionic liquids, tetraalkylammonium bis(2-ethylhexyl)sulfosuccinates, and their ionic liquid–water two-phase properties. *Green Chem.* **2006**, *8*, 349–355. [CrossRef]
104. Cojocaru, O.A.; Bica, K.; Gurau, G.; Narita, A.; McCrary, P.D.; Shamshina, J.L.; Barber, P.S.; Rogers, R.D. Prodrug ionic liquids: Functionalizing neutral active pharmaceutical ingredients to take advantage of the ionic liquid form. *MedChemComm* **2013**, *4*, 559–563. [CrossRef]
105. Zhang, D.; Wang, C.X.; Han, W.; Yang, X.Y.; Qu, Y.; Cui, X.M.; Yang, Y. *Promotion on In Vitro Percutaneous Absorption of Trace Ginsenoside rh1 Using Imidazole Type-Ionic Liquids*; Tianjin Chinese Herbal Medicine Magazine: Tianjin, China, 2014; Volume 45, pp. 2917–2923.
106. Heckenbach, M.E.; Romero, F.N.; Green, M.D.; Halden, R.U. Meta-analysis of ionic liquid literature and toxicology. *Chemosphere* **2016**, *150*, 266–274. [CrossRef] [PubMed]
107. Hough-Troutman, W.L.; Smiglak, M.; Griffin, S.; Reichert, W.M.; Mirska, I.; Jodynis-Liebert, J.; Adamska, T.; Nawrot, J.; Stasiewicz, M.; Rogers, R.D.; et al. Ionic liquids with dual biological function: Sweet and anti-microbial, hydrophobic quaternary ammonium-based salts. *New J. Chem.* **2009**, *33*, 26–33. [CrossRef]
108. Hough, W.L.; Smiglak, M.; Rodriguez, H.; Swatloski, R.P.; Spear, S.K.; Daly, D.T.; Pernak, J.; Grisel, J.E.; Carliss, R.D.; Soutullo, M.D.; et al. The third evolution of ionic liquids: Active pharmaceutical ingredients. *New J. Chem.* **2007**, *31*, 1429–1436. [CrossRef]
109. Bica, K.; Rijksen, C.; Nieuwenhuyzen, M.; Rogers, R.D. In search of pure liquid salt forms of aspirin: Ionic liquid approaches with acetylsalicylic acid and salicylic acid. *Phys. Chem. Chem. Phys.* **2010**, *12*, 2011–2017. [CrossRef] [PubMed]
110. Bica, K.; Rogers, R.D. Confused ionic liquid ions—A "liquification" and dosage strategy for pharmaceutically active salts. *Chem. Commun.* **2010**, *46*, 1215–1217. [CrossRef] [PubMed]

111. Zavgorodnya, O.; Shamshina, J.L.; Mittenthal, M.; McCrary, P.D.; Rachiero, G.P.; Titi, H.M.; Rogers, R.D. Polyethylene glycol derivatization of the non-active ion in active pharmaceutical ingredient ionic liquids enhances transdermal delivery. *New J. Chem.* **2017**, *41*, 1499–1508. [CrossRef]
112. Miwa, Y.; Hamamoto, H.; Ishida, T. Lidocaine self-sacrificially improves the skin permeation of the acidic and poorly water-soluble drug etodolac via its transformation into an ionic liquid. *Eur. J. Pharm. Biopharm.* **2016**, *102*, 92–100. [CrossRef] [PubMed]

MDPI
St. Alban-Anlage 66
4052 Basel
Switzerland
Tel. +41 61 683 77 34
Fax +41 61 302 89 18
www.mdpi.com

Pharmaceutics Editorial Office
E-mail: pharmaceutics@mdpi.com
www.mdpi.com/journal/pharmaceutics

www.ingramcontent.com/pod-product-compliance
Lightning Source LLC
LaVergne TN
LVHW070702170726
843515LV00004B/466

* 9 7 8 3 0 3 6 5 0 0 5 6 0 *